A HISTORY OF
SPACE
EXPLORATION

AND ITS FUTURE...

A HISTORY OF
SPACE
EXPLORATION

Tim Furniss

AND ITS FUTURE...

THE LYONS PRESS
Guilford, Connecticut
An imprint of The Globe Pequot Press

A HISTORY OF SPACE EXPLORATION

Copyright © 2003 by Thalamus Publishing

First Lyons Press edition 2003

The Lyons Press is an imprint of The Globe Pequot Press

10 9 8 7 6 5 4 3 2 1

Printed in Italy

For Thalamus Publishing
Project editor: Neil Williams
Design: Oliver Frey
Four-color separation: Michael Parkinson and Thalamus Studios

ISBN 1-58574-650-9

Library of Congress Cataloging-in-Publication Data is available
on file.

The author accepts responsibility for any factual errors in this
title and acknowledges with thanks the use of the following ref-
erences—in addition to the author's previous works—for access
to statistical data:
Solar System Log, Andrew Wilson; Janes
Skywatching, David Levy; Collins Aeronautics and Astronautics
1915–1960; NASA; Dorling Kindersley Eyewitness Guide to
Astronomy; Phillips Astronomy Encyclopedia; TRW Space
Logs.

Picture Acknowledgments
Bettman/CORBIS: 10; 164, NASA/CORBIS: 4–5.
Stocktrek/CORBIS: 2–3. All other images were supplied by the
Genesis Space Photo Library, which acknowledges with
thanks the co-operation of the public relations and press offices
of the following organizations:
Boeing, British Aerospace, China Great Wall Industry
Corporation. European Space Agency, Grumman Aerospace,
International Launch Services, Lockheed Martin, NASA,
National Commission on Space, NOAA, Novosti,
North American, NPO Energia, Open University, Smithsonain
Institute, Space Telescope Institute, U.S. Army, U.S. Navy,
U.S. Air Force

Page 1: A Space Shuttle is
launched using newly
developed fly-back boosters.

Title pages: View of the
Tarantula nebula.

This page: Space Shuttle
astronaut Bruce McCandless
maneuvers above Earth in his
individual-propulsion suit,
February 1984.

CONTENTS

THE RACE INTO SPACE

I n the 1940s, when space travel was an exciting dream that seemed tantalizingly close, artists produced visionary scenes of the first moon base, space stations looking like bicycle wheels, and men exploring Mars—all predicted to occur before the 21st Century.

It didn't quite turn out that way, but all the same, the first 45 years of the Space Age have seen mankind making its first tentative steps beyond Earth. The 1960s were particularly exciting, with progress being driven at an astounding rate by the intense rivalry of the Cold War. Today, we have the curiosity, desire, and imagination, but no longer the political commitment to drive technology forward.

Our first steps have seen remarkable advances and feats. Before the dawn of the Space Age, it took 149 years for the planets Uranus, Neptune, and Pluto to be discovered by telescope, but with unmanned space probes just 27 years to explore all the planets except Pluto. Asteroids and comets have also been explored at close range. The footsteps of 12 men still lie at six locations on the moon, and 850 lb (385 kg) of moon rock has been brought back to Earth.

In recent decades, space exploration has become space exploitation. Our world is served by countless satellites for communications, meteorological observation, navigation, science, and military uses, and the activity of the sun is monitored daily. Over 400 people have sampled the delights and trials of spaceflight, recently including two wealthy space tourists who have paid tens of millions of dollars for the privilege.

We have dreams of going to Mars and finding life there and elsewhere. Space hotels, Mars bases, and interstellar travel are visions that may one day become reality. But all these will require a special

combination of factors similar to those that ignited the space race.

Compared with the predictions, the reality of space travel so far may seem to be rather an anticlimax, but we have made many astounding discoveries on our first steps into the Universe.

Main picture:
A 1950s vision of a rotating, gravity-inducing space station, with a rocketship and space-walking construction workers alongside.

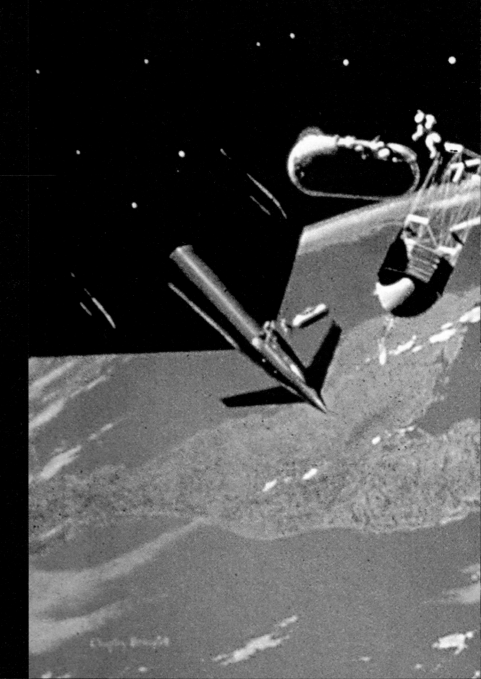

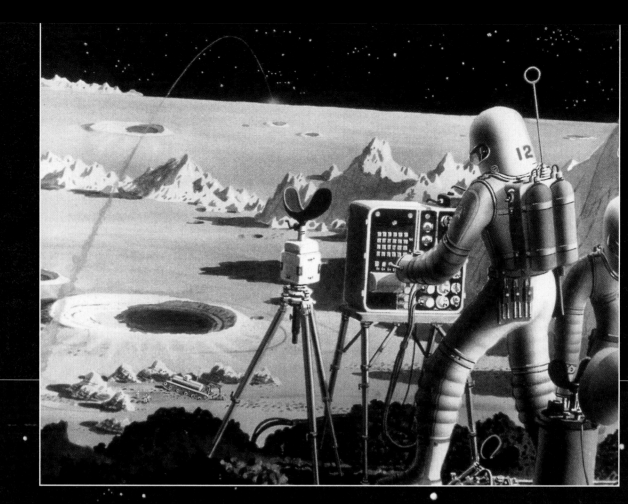

Right: Man on the moon 1950s-style. Notice the spectacular scenery featuring a flat horizon and sharp, craggy mountains.

CHAPTER ONE

ROCKET PIONEERS

Tsiolkovsky and Goddard

For all of history the world had only one satellite—the moon—and space travel was a dream. Then, independently, a Russian schoolteacher and an American scientist made striking developments in rocket propulsion that ushered in the reality of the Space Age.

The teacher, Konstantin Tsiolkovsky, who was born in 1857, is credited as the first person to fully understand the concept of rocket propulsion. In 1883, Tsiolkovsky wrote that a rocket could operate in the vacuum of space because the vehicle moves as a result of the force being expelled from the nozzle. The popular belief among scientists at the time was that a rocket moved as a result of thrust against the air, and therefore it was assumed a rocket could not work in the vacuum of space.

By 1903 Tsiolkovsky was designing a liquid hydrogen-liquid oxygen powered rocket engine, and later experimented with various designs of rockets, including some with stages, which enabled the rocket to shed weight as it headed for space. He realized that a rocket needed to be stabilized in flight, and designed what today is known as a gyroscope. His extraordinary vision included the understanding of attaining orbital flight at a speed of five miles (eight km) per second.

His most famous works were entitled, *Investigating Space with Reaction Control Devices and Cosmic Rocket Planes*, and by 1922, his conceptual work was being put into practice by the Leningrad Gas Dynamics Laboratory (GRL), which eventually developed the first Soviet liquid propellant rocket engine.

Tsiolkovsky died in 1935 but he had laid the foundation of the Space Age. Meanwhile, Robert Goddard was gaining an understanding of rocket propulsion and multistage rockets, and even predicting the use of rockets to explore the planets. Going one step further than Tsiolkovsky, Goddard even theorized on other potential technologies, such as ion propulsion.

Reaching for the sky

Like Tsiolkovsky, Goddard realized that solid propellant gunpowder rockets were not the future, but more efficient liquid propellant vehicles were. He wrote that liquid propellant engines produced far more energy per unit of mass.

They were also more controllable. The quiet and modest Goddard first came into the limelight in 1920 with his famous publication entitled *A Method of Reaching Extreme Altitudes*, which was published by the Smithsonian Institute.

Three years later, Goddard tested a pump-fed liquid oxygen-gasoline engine. Then on March 16, 1926, the great breakthrough occurred. What is regarded as the beginning of the Space Age was heralded by the launch of the world's first liquid fueled rocket, from Auburn, Massachusetts—the Kitty Hawk of space.

The rocket traveled 183 ft (56 m) in 2.5 seconds, reaching a speed of about 62 mph (100 km/h). After flying another vehicle equipped with instruments and a camera, Goddard was awarded funds from the Guggenheim family, and this enabled him to move operations to Roswell, New Mexico, in 1930.

An 11 ft (3.35 m) long rocket was built and flew to an altitude of 1,998 ft (609 m) at a speed of 497 mph (800 km/h), and later another rocket reached a new record of 1.42 miles (2.28 km). These tests were made to improve the stabilization techniques for the rocket design. The first launch using a gyroscope stabilization system was made on April 19, 1932.

Further advances were made during the following 11 years at Roswell, but inexplicably, the U.S. Government showed little interest in Goddard's pioneering work, and consequently the lead that the U.S. had made in rocket technology was somewhat wasted.

Goddard was left working for the U.S. Army and Navy on liquid propellant-assisted take for aircraft. He died in 1945. Nevertheless, Goddard's invention of the first liquid-propelled rocket had ignited the beginning of the Space Age.

The Vengeance Weapon

Germany could have launched the world's first intercontinental ballistic missile in 1945, and would probably have become the first nation to launch a satellite if World War II had not been instigated by the Nazis. However, Germany's lead in space technology came to an abrupt end with its military defeat.

Von Braun designed an upgraded version of the V-2 with a second stage, which could have launched a satellite.

Ironically, it was a World War II German rocket scientist who masterminded America's quest for a moon rocket. While Tsiolkovsky and Goddard were in the midst of their rocket work, a third pioneer was about to emerge—Wernher von Braun. In 1927, the German Hermann Oberth helped to establish the Society for Space Travel (VfR), which successfully developed and fired a liquid oxygen-kerosene rocket from Raketenflugplatz, a rocket firing field outside Berlin. Joining the VfR in 1930 as an enthusiastic 18-year-old was von Braun. By 1932 the VfR was firing larger Repulsor rockets, attracting interest from the German Army, which organized a demonstration firing at Kummersdorf, about 40 miles (65 km) south-east of Berlin.

The German government had been concerned about Oberth's connections with rocket groups in other countries, and his funds were restricted. But Adolf Hitler came to power and increased the VfR's budget almost by a factor of ten, to a value of over 10 million Reichsmarks. The road to space had begun. The VfR effectively became part of the German Army Weapons Department based at Kummersdorf, under the command of Captain

Walter Dornberger, with von Braun as one of his main cohorts.

A liquid oxygen and kerosene engine with 661 lb (300 kg) thrust was developed and fitted to a prototype rocket called the A-1, which failed, but a new vehicle, the A-2, succeeded in 1934, reaching 1.6 miles (2.5 km) altitude. The A-3 was then designed, to include exhaust vanes and fin-mounted rudders.

The U.S. east coast under threat

The German Army established a rocket factory staffed by 12,000 people in Nordhausen and a launch base at Peenemünde on the Baltic Coast. An improved A-5 vehicle was tested from here in 1939, leading to the full development of the next vehicle, the "long distance rocket" called the A-4, to be capable of flying almost 400 miles (640 km) and carrying a payload of 165 lb (75 kg).

The A-4 experienced two failed Peenemünde launches before a flight on October 3, 1942 ,which reached an altitude of over 50 miles (80 km), flying almost 125 miles (200 km) downrange. Dornberger said "today, a spaceship was born." Ironically the spaceship was renamed the V-2 ("V" for Vengeance Weapon) and armed with explosives.

The next V-2 didn't reach space, however, but Paris on September 4, 1944, while two others launched on the same day hit central London and a suburb of the city. Within 13 days another 26 had been launched and a further 2,763 were flown

before the end of the Second World War, with the loss of many thousands of lives.

The V-2 pioneered many technologies. Its engine burned liquid oxygen and a mixture of ethyl alcohol and water, developing a thrust of 55,000 lb (24,950 kg). The propellants were fed into the combustion chamber by pumps driven by a turbine.

It is fortunate that the war ended when it did for the Peenemünde team was working on an even bigger missile, the A-9, an upgraded, winged version of the V-2 which was to be the first intercontinental ballistic missile, and was to be aimed at the east coast of the United States. This made a test flight to an altitude of 56 miles (90 km) with a speed of 2,680 mph (4,320 km/h). Von Braun had plans for A-9 derivatives that would one day fly into space, reaching speeds to enable the first satellite to be placed into Earth orbit. If World War II had not occurred, Germany would surely have become the first space nation.

Left: Nazi Germany's most advanced rocket, the A-9, was the world's first ICBM and would have inflicted destruction on U.S. cities such as New York and Washington if World War II had not ended in May 1945.

Below: Von Braun with a V-2 model in the early 1950s. He joined the German Society for Space Travel in 1930, at the age of 18.

The Peenemünders

In the last days of World War II, the allies were rapidly approaching the top-secret Peenemünde site. As a consequence, in early May 1945 the German SS was given orders to kill the V-2 rocket team. However, von Braun—his arm in a sling after a motorbike accident—escaped from the rocket base along with several colleagues, and eventually surrendered to a Private in the U.S. 44th Division. It was an event that would shape history.

The U.S. Army recovered over 36 tons of documents from the Nordhausen factory but neglected to destroy the plant, and as a consequence valuable V-2 details, hardware, and also personal expertise was acquired by the Soviets.

Under the U.S. Operation Paperclip, von Braun and fellow workers, called the Peenemünders—and components for 60 German V-2s—were taken to the United States, to work at the White Sands Proving Range in New Mexico. Meanwhile, the less fortunate Peenemünders headed for eastern hardship.

Von Braun soon got to work at White Sands, launching the first U.S.-assembled V-2 on April 16, 1946, followed by 11 launches up to the beginning of October, the ninth of which reached a record altitude of 111 miles (180 km).

On October 24, the 13th V-2 made history carrying a camera which made motion picture coverage of the Earth from space, 65 miles (104 km) high. The camera was recovered in the instrument package, which parachuted to Earth.

A V-2 flown for the U.S. Navy returned a camera that had taken the first still photo of the Earth from space. By 1948, the V-2 was flying with an electronic flight control unit and for the first time became a two-stage rocket. The second stage was based on a U.S. Army Corporal missile and the vehicle was renamed the Bumper WAC.

A new launch site

The Corporal had already flown several science missions into near space from White Sands, reaching 50 miles (80 km) on March 22, 1946, the first American rocket to escape the Earth's atmosphere. Another rocket based on V-2 technology was called the Aerobee, which flew many successful sub-orbital science missions, with a first flight on September 25, 1947. A new model, Aerobee-Hi was launched on April 21, 1955.

The Bumper WAC reached a record altitude of 242 miles (390 km) on February 24, 1949, and the following June carried a monkey called Albert 2 who unfortunately died after an unusually hard landing.

A V-2 rocket is prepared for launch from White Sands, New Mexico. The first U.S.-assembled V-2 was launched in April 1946.

A milestone was marked on July 24, 1950 when the Bumper WAC 8 was launched from a new site. This remote, mosquito-infested, swampy sand spit in Florida was chosen because the vehicles could fly safely on longer distances over the Atlantic Ocean. It was called Cape Canaveral. Five days later, another Bumper WAC attained a record speed of Mach 9, nine times the speed of sound. The V-2 program finished at flight 66 on October 29, 1951, with eight Bumper WAC flights, two from Canaveral.

A U.S. derivative of the V-2 was called Hermes and was first launched on May 19, 1950. The program was terminated in 1954. By then the von Braun team had been transferred to Huntsville, Alabama to establish the Redstone Arsenal to develop America's first intermediate range ballistic missile. Meanwhile, the U.S. Navy had developed the Viking rocket—originally called Neptune. The rocket illustrated the progress that was being made in lightweight materials, integral tanks, gimballing engines for steering, and exposing scientific instruments directly to space. The Viking flew 11 missions from White Sands and one from a U.S. Navy ship between May 3, 1949 and February 4, 1955.

The Viking reached an altitude of 157 miles (252 km) in 1954. The vehicle made many scientific measurements at high altitude, including atmospheric winds and density. The model formed the basis of a new rocket called Vanguard, with which the U.S. Navy hoped to launch the first satellite into space. However, in the secretive Soviet Union, another team of engineers had different ideas…

Left: The U.S. Viking sounding rocket was the precursor of the Vanguard, which was planned to carry the first U.S. satellite into orbit.

Below: This historic first launch from Cape Canaveral, Florida, in July 1950 featured a V-2 with a Corporal second stage.

The Chief Designer

In the early 1930s, as the Leningrad Gas Dynamics Laboratory (GDL) was working on the Soviet Union's first liquid propellant rocket motor, the ORM 1, the Group for the Study of Reaction Propulsion (GIRD) was being established in Moscow. Its chief engineer was Sergei Korolov, a test pilot and aircraft and glider designer, who joined GIRD in 1932.

GIRD 9, which was to become an anti-aircraft missile, was launched on August 17, 1933, reaching a height of about one mile, while GIRD 10 was a liquid propellant booster which reached three miles (4.9 km) on November 25, 1933.

GDL and GIRD were then merged into the Scientific Rocket Research Institute (RNII), and in 1939, its first two-stage rocket was launched. In 1945 its work was infused with German technology when General Rokossovsky of the Russian Army entered Peenemünde and plundered what V-2 resources there were, including manpower. For the next two years, the Soviets produced more V-2s than Germany had in the whole of the War.

Korolov was assigned the job of producing the Russian version of the V-2 and became known as the Chief Designer, becoming anonymous to the outside world. A new Russian version of the V-2, called the T-1, was then produced, followed by an astrophysical rocket, capable of flying to altitudes of over 60 miles (100 km), carrying a suite of science instruments weighing over 170 lb (77 kg).

The T-2 followed, raising the capability to 120 miles (200 km) and 280 lb (127 kg), becoming the Soviet Union's intermediate range ballistic missile (IRBM), the equivalent of the U.S.'s planned Redstone.

Beginnings of the Space Race

In 1955, the Soviet Union agreed to participate in the International Geophysical Year (IGY) in 1957–58, flying instrumented rockets, including the T-2, into the upper atmosphere. The Soviets also announced their plan to launch an Earth satellite to contribute to the IGY.

Korolov was already working on the satellite's launcher—the T-3, which was to eventually prove its capability by flying missions to altitudes of over

Above: Sergei Korolov, who became known anonymously as the "Chief Designer," was the architect of the Soviet Union's space program.

130 miles (209 km), capable of carrying a payload of over two tons. The T-3—known as the R-7—was in fact the first Soviet intercontinental ballistic missile (ICBM), capable of delivering a nuclear warhead onto any target in America.

In 1955, Korolov moved to a new base accessible by railroad in the midst of the flat, remote, and inhospitable Kazakhstan steppes, and built a launchpad over an old mine pit which would serve as the flame trench for the new missile. Known as the SS-6, or Sapwood, by U.S. intelligence chiefs, the booster was photographed on its launchpad at Tyuratam, Kazakhstan, by a U-2 spyplane. It was first launched in 1956 and made over 50 flights, with long range testing beginning in May 1957 with the first of eight missions, ending in August with a flight of over 10,000 miles

(16,000 km). The Soviets announced that they had an operational ICBM. America was also belatedly planning to develop an ICBM, called the Atlas, but it was falling behind.

The U.S. chose to disregard the Soviet Union's previous announcement of its plans to launch a satellite in 1955, and blissfully continued preparations to launch their IGY satellite, called Vanguard, aboard a purely civilian rocket, with no connection with military missile technology.

The rocket designed to carry the satellite was also called Vanguard and was a derivative of the Viking rocket operated by the U.S. Navy.

The U.S. Army Redstone team under von Braun proposed and designed a version of the Redstone that could launch a satellite, but was told to back off from the civilian and scientific IGY plan because the rocket was based on a missile and would be bad public relations.

The inevitable happened. On October 4, 1957, Korolov launched his R-7 derivative with Sputnik in its nose cone—and America was overcome with shock and trepidation.

Right: A Russian V-2 derivative astrophysical rocket called the T-1 was capable of reaching 60 miles altitude.

Left: The world's first intercontinental ballistic missile was the Soviet Union's R-7, seen here at Tyuratam, Kazakhstan.

15

CHAPTER TWO

THE FIRST STEPS

Sounds That Shocked the West

The design of Sergei Korolov's R-7 intercontinental ballistic missile (ICBM) was brilliantly simple. Korolov basically stacked five "first" stages together to create a rocket with sufficient thrust to reach orbit. All he had to do was to put a nosecone on the top with a small satellite inside and the Soviets would be first in space. Sputnik was ready.

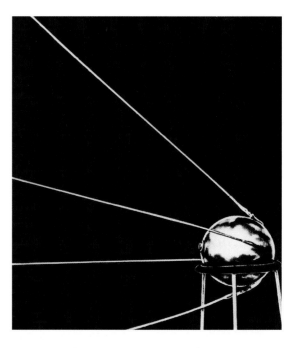

A s the International Geophysical Year (IGY) approached, all three U.S. military services proposed a satellite project to contribute to the worldwide science project. President Dwight D. Eisenhower vetoed the Army Bomarc winged missile and the Air Force Redstone missile-based Jupiter C teams, and opted for the U.S. Navy Vanguard, as it was to be launched on a "peaceful" civilian-developed rocket derived from the Viking. The Soviet Union's announcement of its IGY satellite plan was largely ignored.

Above: The world's first artificial Earth satellite, Sputnik 1, was launched on October 4, 1957, marking the start of the Space Age.

An American satellite could have been placed into orbit on August 7, 1957 had a Jupiter C, flying to a record altitude of 597 miles (960 km), been equipped with an upper stage. The furious and frustrated von Braun had been forbidden to do so by Eisenhower.

Below: Sputnik 2 carried Laika, the first animal in orbit. She was killed by extreme temperatures in her tiny compartment.

The Vanguard project was languishing on the launch pad. Korolov probably couldn't believe his luck. He was ready. In the dead of night on October 5, at Tyuratam—October 4 in Moscow—the modified R-7 missile was launched and reached space, 134 miles (215 km) up, traveling at a speed of 4.96 miles (7.99 km) per second. A silver sphere about two feet (60 cm) in diameter was deployed from the upper stage, weighing 184 lb (83 kg), with four 9.8 ft (3 m) long antennas. The satellite later reached a peak altitude of 583 miles (939 km) and crossed the equator at an inclination of 65.1°. One orbit took 96 minutes.

Three objects were now in orbit, Sputnik 1, a rocket stage, and a payload fairing, the latter two ironically being the first pieces of space debris. Radio Moscow announced the news of Sputnik's launch to a shocked world, and to make matters worse, the satellite's two radio transmitters released continuous tones: "bleep… bleep… bleep…," the sounds that haunted America. The Soviet Union—mistakenly regarded in the West as a technologically backward nation—not only had an ICBM capable of launching a nuclear warhead but was also first in space. For America—and the West in general—in the midst of the Cold War, living with the very real threat of nuclear war at any moment, Sputnik was a great shock. But more was to come, and different sounds were to haunt the West.

The first death in space

On November 3, 1957 another Korolov R-7 derivative headed for the skies, placing a conical shaped craft, Sputnik 2, into orbit. Remaining attached to the core stage, the satellite weighed over seven tons. Inside a pressurized container in the spacecraft was the first living creature to orbit the Earth (although many animals had been into space on sub-orbital flights). The female husky dog called Laika was in space and on a one-way journey. At this early stage of the Space Age there were no re-entry and recovery capabilities. The dog's barking was heard in some transmissions while other instruments on the satellite sent back data on solar and cosmic radiation and about a possible radiation belt around the Earth.

Laika barked and moved about in the tight confines, and became highly agitated as the temperature rose to 40°C. The unbearable heat eventually killed her, well before the oxygen ran out. The satellite later re-entered the Earth's atmosphere, on April 14, 1958.

The U.S. could have launched the first satellite in 1957 using an upgraded version of the Redstone intermediate range ballistic missile, but Eisenhower vetoed the plan.

Kaputniks and Flopniks

December 6, 1957 can be regarded as America's space nadir. With over seven tons of Soviet hardware orbiting above, the U.S. prepared to launch its answer: the long-awaited Vanguard satellite—all 2.9 lbs (1.35 kg) of it.

Far right: Celebrating the launch of Explorer 1, the first U.S. satellite, and holding a model of the spacecraft are (left to right): William Pickering of the Jet Propulsion Laboratory which built the satellite, James van Allen, who designed the radiation detectors, and Wernher von Braun, the Jupiter C director.

Main picture: The Vanguard rocket falls back onto the launch pad. The satellite survived and is today exhibited at the National Air and Space Museum in Washington, DC.

Below: America's answer to Sputnik—the tiny Vanguard satellite which never made orbit on December 6, 1957.

The launch was originally planned as a low-key test flight with a small test article satellite which may or may not have made orbit, as engineers felt that they were not yet ready. The Vanguard launcher had not even been fully-tested in its complete configuration.

The U.S. Navy-led team were rather angry and surprised by the announcement from the White House that America's first satellite was about to be launched from Cape Canaveral. They thought that a confidential briefing was going to be just that.

Covered live on TV in the U.S., the countdown reached zero, and the slender, pencil-like Vanguard rose from the launch pad, reaching four feet before falling back and exploding in a black and orange ball of fire. Ironically, its tiny satellite fell away largely unscathed and today is exhibited at the National Air and Space Museum in Washington, DC.

The press had a field day, coming up with inspiring headlines such as "Kaputnik" and "Flopnik." U.S. Senator Lyndon B. Johnson spoke for the nation when he said the failure had been "most humiliating." The exasperated von Braun was given a free hand and told to fly his Jupiter C—and the recovery began. However, the Kaputnik debacle would haunt America for years and only serve to perpetuate the Cold War Space Race.

The 28.7 lb (13 kg) Explorer 1 was mounted on a little rocket stage on top of the von Braun Jupiter C at Cape Canaveral and successfully entered orbit on February 1, 1958. Reaching 1,583 miles (2,548 km) at the highest point—the apogee—of its orbit, the satellite's Geiger counters designed to measure cosmic rays became saturated with other radiation. It became apparent that the Earth had a radiation belt surrounding it.

Scientific applications

Explorer had proved the value of satellites, and the program of the same name went on to greater things. Thirteen more successful Explorer satellites launched up to the end of 1962 continued observations of the radiation belt, studied magnetic fields, solar flares, the ionosphere, atmospheric density, gamma rays, and micrometeorites. The Explorer series continues today.

A Vanguard satellite was finally launched on March 17, 1958 after another launch failure, and

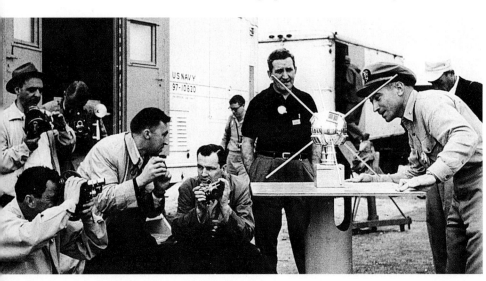

demonstrated the use of solar cells to derive electricity for a spacecraft, rather than a limiting battery. An Explorer satellite, No 6, launched in 1959, flew an array of solar cells on four paddlewheel panels attached by arms to the body of the spherical spacecraft, rather than on the craft itself as in the case of Vanguard 1.

Two further Vanguard satellites were launched but there were other launch failures. The program was not entirely a success. There were 14 launches in all, including three successful test flights but which did not make orbit, the Kaputnik debacle, another failure, the success of Vanguard 1, four more failures, Vanguard 2, two further failures, and Vanguard 3. Indeed, launch failures featured highly in the early days of the Space Age. Since American launches were not secret, the nation bore its losses publicly while Soviet launch failures remained secret. Consequently, when Sputnik 3 was launched in May 1958 with a dedicated geophysics and space physics payload, the impression was that the USSR had made three successful launches in a row. But there had been one failure the

previous month. Indeed, the combined Soviet and U.S. launch record for 1957–59, including launches from Earth into interplanetary space, shows 27 failures out of 51 launches. Of these, there were 20 failures by the U.S. with 18 successes, and seven Soviet failures and six successes.

The most spectacular U.S. failure was the Juno 2 rocket, which turned tail on the launchpad in 1959, destroying a Beacon satellite. This was during a record sequence of seven consecutive failed launches between April 13 and July 18, 1959, two by the Soviets and five by the U.S.

By 1960, as the Soviet Union exclusively concentrated on lunar and planetary launches and preparations for manned spaceflights, with limited success, the U.S. was leading the way in demonstrating space applications. Today we take communications, weather forecasting and Earth observation, navigation and military satellites for granted, but the pathfinders for these applications were the early satellites of the Space Age.

Early Space Science

In the frenetic excitement of the early days of the Space Race it was often forgotten that the first satellites were supposed to be launched as part of the International Geophysical Year. Most U.S. and Soviet satellites had a serious scientific purpose.

The IGY was a period between July 1957 and December 1958 when many countries co-operated in a study of the effects of solar activity on the Earth using sensors, not just on satellites but on balloons, remote ground-sites, and other media. According to the IGY concept, the first spacecraft to be launched into orbit should be science satellites, and Sputnik 1 and 2 did indeed carry sensors to measure solar ultraviolet and X-ray radiation and cosmic rays.

In 1958, Explorer 1 contributed to the greatest discovery. A cosmic ray detector on the pencil-shaped craft was saturated with a flux of charged particles. The detector, designed by a team led by James Van Allen, had discovered that highly energetic protons and electrons were trapped in the Earth's magnetic field—there was a radiation "belt" about 600 miles (965 km) above the Earth.

Later in 1958, the Pioneer 3 spacecraft (which was launched with the moon as its target but actually reached about 63,958 miles [102,333 km] before falling back to Earth) detected a second and higher radiation belt encircling the Earth. Electrons and protons were surrounding the Earth in a continuous cloud as far away as ten Earth radii and oscillating between the northern and southern hemispheres along magnetic field lines.

Near-space environment

In 1959, three Soviet Luna spacecraft and America's Pioneer 4 made a new discovery: the solar wind, a stream of radiation from the sun that interacts with the Earth's magnetic field, causing aurorae.

The sun was a far more dynamic star than was first imagined. Later satellites showed that ionized particles in the solar wind were blowing the Earth's magnetic field into a teardrop shape, forming the magnetosphere.

Satellites provided a map of the magnetosphere. The solar wind hits the magnetic field on the sunward side of the Earth and is diverted into a shock wave called the bow shock. Inside the bow shock is a turbulent ionized region called the magnetosheath or magnetopause. On the opposite side of the Earth, the magnetosphere is drawn out to a great distance like the wake of a ship.

Over the first five years of the Space Age several science satellites were launched to explore other areas of the Earth's near-space environment and further afield. Varied areas of space science were pursued and observed by satellite, especially the U.S. Explorer series, which studied the ionosphere, atmospheric density, geomagnetic field, gamma rays, and meteoroids.

Explorer 3 transmitted stored data on cosmic ray bombardment. Cosmic rays or radiation are very penetrating atomic particles which bombard the Earth from deep space, some emanating from supernova explosions light years away.

Explorer 6 measured the behavior of radio waves in the Earth's ionosphere and helped map the magnetic field. Explorer 15 studied the potential danger to spacecraft posed by meteoroids or micrometeorites, which hit the atmosphere regularly at speeds of up to 45 miles per second (72 km/s).

The Soviet Union also launched similar satellites but not as many or varied as the Explorers. The flight of Sputnik 3, launched on May 15, 1958, was one of the most scientifically-packed early missions. The 11 ft (3.35 m) long, 2,914 lb (1,322 kg) Sputnik 3 carried a magnetometer, light intensifiers for recording corpuscular radiation from the sun, and photon recording devices (a photon is the smallest packet or quantum of light energy). The satellite also measured ions, or atoms, cosmic rays, and micrometeorites. The dynamics of the sun also came under close scrutiny with the launches of purpose-built observatories, such as the American Orbiting Solar Observatory.

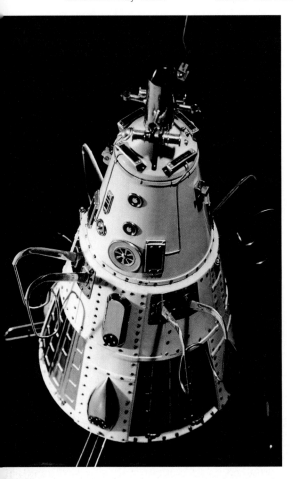

Below: The Soviet Union's Sputnik 3, the nation's first dedicated science satellite, was launched in May 1958.

Inset: Explorer 15, launched in October 1962, carried an array of sensors to measure the radiation enveloping the Earth.

Main picture: A Thor Able booster launches Explorer 6 from Cape Canaveral in August 1959. It was called the "paddlewheel" satellite, the first to carry solar arrays on extendible booms.

Exploring Exploitation

Communications satellites are extremely common today, as are navigation satellites used by car drivers, science satellites monitoring the sun's "weather," and classified military satellites taking images of the Earth with the ability to see objects less than one foot (30 cm) across. In just 45 years space technology has revolutionized the way we live.

Below: Telstar 1, the first TV broadcasting satellite, was launched in July 1962 and transmitted the first live (but brief) TV programs between the U.S. and Europe.

The space industry today is supported by an international fleet of commercial launchers, while the ground segment industry, including receivers, is growing at an extraordinary rate. The potential for space exploitation was

demonstrated by the first satellites. The first active communications satellite was called Telstar, which was launched by the U.S. in July 1962 and allowed live TV from the U.S. to be seen in Europe for the first time. The 2.6 ft (0.8 m) diameter, 170 lb (77 kg) satellite carried just one transponder and was placed into a low Earth orbit, so could be used only when it was within line-of-sight with two communications receivers. Telstar began the revolution in communications that later featured a fleet of geostationary satellites (i.e. orbiting at a fixed spot above the Earth).

The first weather satellite was called Tiros and was launched by the U.S. in April 1960. The name of the satellite was an abbreviation of Television and Infra Red Observation Satellite. The 3.4 ft (106 cm) diameter craft weighed 263 lb (119 kg). Its cylindrical body was covered with 9,260 silicon solar cells which generated electricity from the energy of the sun. Some of the earlier satellites had been battery-powered, so their operational lifetimes were short.

Tiros's wide-angle and narrow-angle high resolution vidicon cameras were equipped with magnetic tape recorders which could store 32 Earth images before transmitting them to a ground station. The images illustrated the potential value of meteorological satellites and provided very useful early-warning predictions of hurricanes.

Spies in space

April 1960 marked another milestone in the first applications of space, with the launch of the U.S. Navy Transit 1B (the launch of 1A failed). This 267 lb (121 kg) spacecraft was the precursor of a planned initial system of three navigation satellites which would provide positioning accuracy of about 500 ft (150 m) for ships and submarines, using trigonometric measurements. Again, at the time this was a revolutionary use of space.

The military application of space was not lost on the U.S., which soon developed a recoverable camera-equipped spy satellite. The U.S. Air Force project was called Discoverer and was flown under

Discoverer was not a very successful program and it was not until August 1960 that a Corona satellite, Discoverer 13, was recovered—after a splashdown. It was the first satellite to come back from space intact. The method had been proven but no film was carried on this mission. The long-awaited film came back on Discoverer 14—successfully snared by a C-119—and when developed revealed details of a military airfield in the Soviet Union as well as other locations. The military application had been successfully demonstrated.

Left: Tiros 1, launched in April 1960, was the first dedicated weather satellite.

Below: The Discoverer 13 re-entry capsule, pictured being prepared for flight, was the object to be recovered from orbit in August 1960.

the guise of a science program, but was known by the CIA as Corona. The satellite was based on the second stage of a Thor Agena booster launched from a new site at Vandenberg Air Force Base, California, which was suitable to launch craft into orbits around the poles of the Earth.

At the tip of the Agena stage was a 300 lb (136 kg) capsule equipped with a camera capable of taking images of the Earth with a resolution of 32 ft (10 m). At the end of the mission, the capsule was released from the upper stage and plunged into the Earth's atmosphere, and would have burned up had it not been for its heatshield. As it descended toward the Pacific Ocean, it was snared by a net-like structure trailed from the back of a C-119 aircraft.

At its inception in 1959

Reaching for the Moon

The moon was the obvious first target for exploration, but the first attempts to reach it were made in undue haste. Stretching 1950s technology to the limit, it was plain that there would be many failures in the quest to reach Earth's nearest neighbor in space.

In July 1958, just nine months after Sputnik 1, the Soviet Union attempted to send an 859 lb (390 kg) sphere to hit the moon. The launch failed—a fact only to come to light decades later. Six other Soviet moon failures were unpublicized and only revealed years later.

The U.S. launched Pioneer 1A to the moon in August 1958. The 83 lb (38 kg) spacecraft was fitted with a retro-rocket and was aimed to orbit the moon and take the first images of its surface using a crude imaging system. It was extraordinarily ambitious, after just four successful U.S. satellite launches and five failures. Unfortunately, the Thor Able rocket exploded at T+77secs after launch from Cape Canaveral.

After another Soviet Luna failure, two other potential lunar orbiters, Pioneer 1B and 2 were launched in 1958 but did not achieve enough velocity and fell back to Earth. Pioneer 1B reached a record distance of 70,717 miles (113,854 km).

Another Luna failed, before Pioneer 3 was launched. This smaller 12.94 lb (5.87 kg) craft, aiming for a lunar-fly by, reached 63,958 miles (102,333 km) before falling back to Earth (*see page 20*). The Soviet Union at last made a successful launch of a Luna craft—which of course they called No 1, despite the earlier failures—on January 2, 1959. Again the aim was to crash-land a capsule on the moon. However, the craft missed the moon by about 3,725 miles (5,995 km) and entered solar orbit, becoming the first artificial planet, a space feat that was claimed eagerly by the Soviets.

Another spacecraft entered solar orbit soon afterward, this time deliberately missing the moon on a lunar-fly by at a distance of 37,500 miles (60,000 km). It was called Pioneer 4. These lunar flights were certainly catching the interest of a general public already excited by the Space Age.

Another lunar impact attempt failed before Luna 2 was launched on September 12, 1959, hitting on the moon the following day close to the crater Archimedes, the event marked by the sudden end to the signals from the spacecraft, which was transmitting measurements from magnetometer and micrometeoroid and radiation detectors. The flight was an historic one: the first contact with the moon by a craft from Earth.

Below: America's Pioneer 1, the first moon probe, was aiming to become the first lunar orbiter. This extraordinarily ambitious mission ended soon after launch when the Thor Able booster exploded in August 1958.

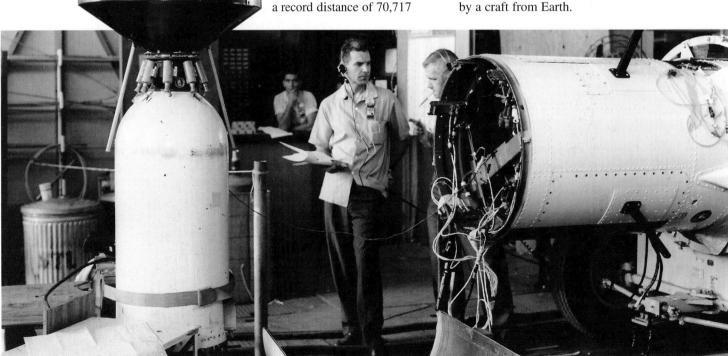

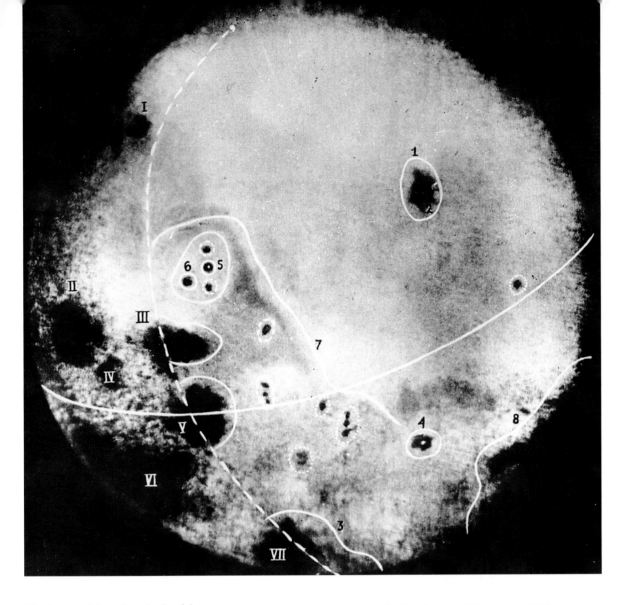

Photographing the dark side

This was soon followed by Luna 3, which went one better in October 1959. Luna 3 was not a lunar lander but designed to enter a deep Earth orbit that would fly around the far side of the moon. It was a tremendous technological achievement. The 612 lb (278 kg) craft was equipped with an extraordinary and ingenious camera system, which involved 35 mm film being developed, fixed, and dried by an on-board processor, scanned by a light beam, and transmitted to Earth rather like a facsimile machine. And it worked.

As Luna 3 passed around the far side at a distance of 40,750 miles (65,200 km) from the moon, the camera system took 29 exposures in 40 minutes. The images were developed and transmitted, revealing 70% of the previously-unseen far side. Although the images were not of great quality, they revealed a far side far more cratered than the near side. For the first time in human history, mankind saw the "dark side" of the moon.

Remarkably, the next fully successful lunar mission was the U.S.'s Ranger 7—not launched until July 1964! Between Luna 3 and Ranger 7 there were 14 moon mission failures.

Below: Pioneer 3 only reached 63,350 miles en route to a planned fly-by of the moon.

CHAPTER THREE

MAN IN SPACE

The Pathfinders

The Russian dog Laika was the first living creature to enter Earth orbit, but she had many sub-orbital predecessors—many of whom also died. Manned spaceflights could not be contemplated until the effects of space travel on animals were known.

Below: The space monkey Gordo was launched on a sub-orbital spaceflight by a Jupiter missile in December 1958, but died when his recoverable capsule sank.

The first steps in space physiology and biological testing were made at Wright Field, Dayton, Ohio in 1935, and later at the U.S. Air Force Aeromedical Laboratory which pioneered the first flights of animals into space.

Albert 1, the first animal known to have been launched on a rocket, flew in 1948. The anesthetized rhesus macaque was launched in the nosecone of a V-2 booster from White Sands, New Mexico. He died when the nosecone impacted at the end of the flight when the parachute failed. This fate befell three other monkeys—including Albert 2, who reached 83 miles (133 km) in June 1949, and two others in 1950. The fifth and final Aeromedical Laboratory V-2 animal mission was launched in August 1950, carrying a non-anesthetized mouse which was photographed by a camera—and survived the landing.

A series of three Aerobee rocket tests began in April 1950, launching from Holloman Air Force Base, New Mexico. These carried mice and primates on flights which ended in 1952 and included a flight to 50 miles (80 km) on September 20, 1951, at a speed of 3,200 mph (5,150 km/h), providing the passengers—a monkey named Yorick and 11 mice—with two minutes of weightlessness. The animals were recovered and were the first to survive the program. The final Aerobee flight, on May 22, 1953, carried two mice called Mildred and Albert and two Philippine monkeys called Patricia and Mike, seated in different positions and recorded by a video camera to watch the effects of acceleration and weightlessness.

The U.S. Air Force also flew animal flights aboard converted Jupiter missiles, in 1958–59. A squirrel monkey, Gordo, was launched in December 1958 but died when his capsule sank after the sub-orbital flight, while Able and Baker reached a record 300 miles (483 km) on May 28, 1959 and were recovered, although Able died from the effects of anesthesia during the removal of electrodes.

Animals in the front line

The NASA Mercury program used chimpanzees for a series of test flights, including sub-orbital and orbital flights by Ham and Enos respectively. Earlier, rhesus monkeys Sam and Miss Sam flew on Little Joe rockets in Mercury systems test missions in 1959–60. Sam reached 55 miles (88 km) while Miss Sam was propelled by the Mercury escape system to nine miles (14 km).

In 1949, the Soviet Union developed a geophysical rocket known as the Mitio that was capable of flying to 120 miles altitude (193 km). By 1951 the Soviets had flown the first Mitio biological missions from Kapustin Yar, north of the Caspian Sea, with dogs in recoverable nosecones. Laika, the first animal to enter orbit, was launched by the Soviet Union aboard Sputnik 2 in November 1957 (*see p.16–17*).

As they were preparing for manned flights, the Soviets tested the Vostok ejection seat on high

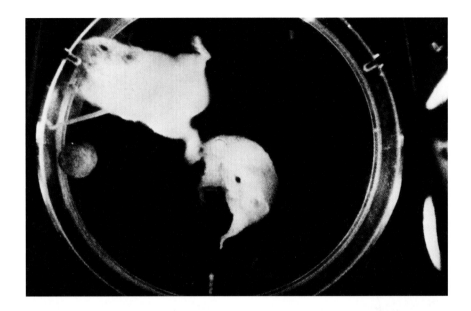

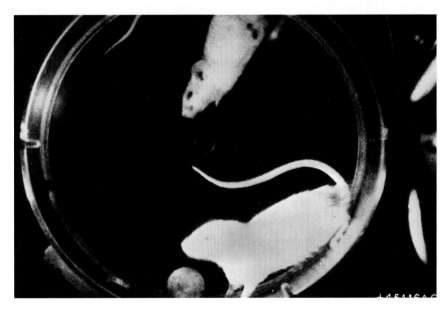

altitude rocket flights. One of the flights carried two dogs, Albina and Tsyganka, dressed in pressure suits, to an altitude of 53 miles (85 km). Other dogs reached 300 miles (483 km).

Later in the space program, many types of animals were launched into space for research purposes, including rats on Apollo 17, frogs, tortoises, an oyster toadfish, crickets—and even a cat, Felix. He was launched on a French Veronique rocket from Algeria on October 18, 1963, reaching an altitude of 120 miles (193 km), and was recovered safely, which can't be said of the next feline passenger who was killed just six days later.

Above: Aerobee suborbital spaceflights in 1950–52 carried mice and primates. Here mice are seen floating in weightlessness.

Left: A V-2 launched from White Sands flew the first animal spaceflight in 1948, carrying a monkey called Albert 1 who died on impact.

The First Rocket Planes

Futuristic impressions of orbital space stations produced in the 1940s–50s featured airplane-like rocket ships shuttling to and from orbit. The first steps to that logical vision came in 1944 when the U.S. Congress approved a "rocket research airplane program." It would be managed by the military services and the National Advisory Committee for Aeronautics (NACA).

Right: Chuck Yeager, the first man to fly faster than the speed of sound, in an X-1 rocket plane on October 14, 1947.

Below: The X-1 rocket plane reached a record altitude of 69,000 ft in 1949, piloted by Frank Everest.

The Bell company was awarded a contract by the U.S. Air Force to produce a piloted transonic aircraft, called the XS-1, powered by a liquid propellant XLR-11 engine produced by Reaction Control Motors, Inc. The engine used liquid oxygen oxidizer and a mixture of ethyl alcohol and distilled water as fuel. The XS-1, later known as the X-1, would be dropped from a B-29 bomber at 30,000ft (9,150 m). Bell test pilot Chalmers Goodlin flew 12 glide flights from a Florida site before the operation moved to Muroc Air Base, in the Mojave Desert, California. The plane made 20 rocket-powered missions using a four-chambered XLR-engine in the tail of the craft, which was piloted by Goodlin and Alvin Johnson from December 1946 to June 1947, before the X-1 was turned over to three designated military pilots, much to the chagrin of the first pilots. The Air Force pilots included Captain Chuck Yeager, who flew into the history books on his 19th mission, on October 14, 1947, by breaking the sound barrier at Mach 1.06. He named his X-1 "Glamorous Glennis," after his wife.

NACA pilots Herbert Hoover and Howard Lilly soon followed while an Air Force pilot, Colonel Frank Everest, reached a record altitude of 69,000 ft (18,288 m). Everest had a chance to use a new T-1 pressure suit, which saved his life.

Further improved models of the X-1 were flown with the aim of reaching Mach 2. The Douglas Skyrocket—based on the jet-powered Skystreak—broke the sound barrier in 1951 and in 1953 went on to reach an altitude of 83,235 ft (25,380 m). On November 20, the same year, pilot Scott Crossfield took it past Mach 2.

To the edge of space

By this time, the Air Force had developed the upgraded X-2 in a bid to reach Mach 3 and altitudes of 130,000 ft (39,650 m). The X-2 was powered by an upgraded XLR-25 throttleable engine and an improved airframe including heat-resistant nickel alloy.

After 11 flights, the X-2 made two flights into history. Air Force Captain Ivan Kincheloe flew the plane to a height of 126,000 ft (38,430 m) on September 7, 1956, and was dubbed the "first spaceman." The handsome Kincheloe had the classic test pilot image, which was perpetuated when he was selected to test-fly the new X-15 rocket plane, which was to fly into space. Kincheloe was, however, tragically killed in an F-104 jet plane accident in 1958 just before he was due to fly the X-15.

The 13th, last and infamous X-2 flight was designated as the one to reach Mach 3. Its pilot was Milburn Apt. On September 27, 1956, Apt flew the X-2, breaking Mach 3, but the plane went out of control and the G-forces built up so much that it was impossible for the pilot to initiate an emergency escape.

Edwards Air Force Base, California (named after an Air Force pilot who lost his life testing an aircraft) became synonymous with the reach for space, in an age when reaching orbit—by airplane—seemed just around the corner. It was considered logical that the way to reach space routinely and cheaply was to follow the spaceplane route, and a new craft arrived in 1958—the one that would, it was hoped, open the road to regular and practical space travel. This was the X-15, and is featured on page 40.

Above: The Skyrocket was the first to reach a speed of Mach 2, in November 1953, piloted by Scott Crossfield.

29

The "Korabl Sputnik"

Soviet engineers began designing a manned spacecraft in 1956 and had their eyes fixed firmly on the moon—even before Sputnik launched the Space Age a year later. After the success of Sputnik the propaganda value of the first manned spaceflight was obvious.

I n 1959, Sergei Korolov finalized the design of the "Korabl Sputnik," and a year later 20 pilots were selected as members of the first cosmonaut group. Many would become household names while others were not so fortunate, including Yuri Bondarenko who died from burns after an oxygen chamber test went badly wrong in 1961.

The Korabl Sputnik became known as Vostok.

Below: The design of the Vostok spacecraft that carried Yuri Gagarin in 1961 was not revealed for several years, and consisted of a basic spherical recoverable capsule.

The 10,430 lb (4.73 tonne) Vostok spacecraft was 14.5 ft (4.4 m) long and 8 ft (2.43 m) in diameter. The cosmonaut flew in a spherical descent module which weighed 5,425 lb (2.46 tonnes) with a diameter of 7.5 ft (2.3 m).

The seat of the spacecraft was an ejection seat to allow the cosmonaut to exit the craft which, even under a parachute, would impact at a fatal speed of 34 feet per second (10m/s). Landing separately under a parachute, the cosmonaut landed at about 16 ft per second (5m/s). The ejection seat would also be used in the event of a launch failure.

A Vostok flight was very simple and the

cosmonaut was basically a passenger rather than a test pilot. The cabin was fitted with a food locker, radio, experiment cabinet, and a porthole with an optical orientation indicator. There was enough room in the cockpit for the cosmonaut to undo straps and float a little upward. The module also had external command, control, and communications antennas. The all-around outer heatshield of the capsule was ablative and burned away during re-entry, absorbing the heat.

Beneath the descent-flight module was an instrument module. This was attached to the descent module by umbilicals and also used four metal straps which extended from the instrument module around the descent module. The 5,000 lb (2,270 kg) instrument section was 7.4 ft (2.25 m) long and 8 ft (2.43 m) at its maximum diameter, and provided the oxygen and nitrogen for the cosmonaut's life support system. At its conical base was a vital retro rocket to slow the craft down for re-entry.

Design for survival

The Vostoks were launched into orbits low enough to ensure that gravity and atmospheric drag would cause a natural re-entry within ten days if the retro rocket happened to fail. The 3,560 lb (1,610 kg) thrust rocket used hypergolic nitrogen tetroxide oxidizer and an amine-based fuel which ignited spontaneously, not requiring an ignition source. It fired for 45 seconds, reducing the orbital velocity by about 500 ft per second (155m/s).

Several test flights were made before a Vostok was entrusted with a human, and it was fortunate that these tests were carried out. There were three failed launches in 1959–60 before the first Vostok was launched into orbit from Baikonur on May 15, 1960 and was called Sputnik 4 basically as a cover. Aboard was a "cosmonaut" dummy dressed in a spacesuit. The spacecraft was wrongly aligned at retrofire and instead of returning to Earth, Sputnik 4 went into a higher orbit. A cosmonaut would have died.

A Vostok was destroyed when a launch ended in an explosion, killing two dogs, Chaika and Lisichka. Sputnik 5 followed in August 1960 and carried the dogs Strelka and Belka, who had the

distinction of being the first animals to be recovered from orbit after a safe retrofire and landing.

Sputnik 6's Pchelka and Mushka weren't so fortunate, being burned up in the atmosphere during a botched re-entry. Two other dogs, Damka and Krasavka were safely recovered after a launch failure later in the same month.

Sputniks 7 and 8 were launched in March 1961 as the final rehearsals of a planned one-orbit flight by the first cosmonaut, carrying the dogs Chemushka and Zvezdochka. All was ready for the first Soviet manned spaceflight.

Above: Strelka and Belka, the first living beings to be recovered from orbit, in August 1960, flew on a Vostok test flight under the guise of Sputnik 5.

Left: The Vostok cosmonauts could not land in the capsule so had to eject at high altitude, another fact not revealed at the time of Gagarin's flight.

Eastern Propaganda

April 12, 1961 was surely one of the greatest days in space exploration. The Soviet Union launched Air Force Captain Yuri Gagarin into orbit aboard a spacecraft called Vostok 1. The entire world was captivated with the Soviet success.

The U.S. had been beaten once more, and it was like Sputnik 1 all over again. The worldwide reaction and acclaim for the Soviets was the last straw. America's official reaction would be given on May 25, 1961 with an historic speech to Congress by the President.

However, what was not realized at the time was that Gagarin had merely been a passenger in his craft, was almost killed during re-entry, and did not even land in his spacecraft. Vostok 1's retro rocket fired for 40 seconds but the instrument section didn't separate from Gagarin's spherical capsule. The spacecraft entered a spin and Gagarin saw a reddish light out of his window as re-entry began, to the accompaniment of crackling noises.

The spacecraft was uncontrollable and close to disintegration. G-forces built up to a level of 10G.

Below: Yuri Gagarin, the first man in space, was killed in an air crash in 1968 while in training for a Soyuz mission.

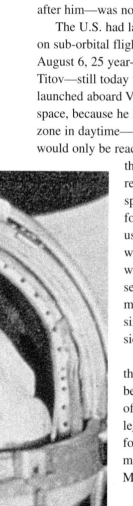

Fortunately, the heat severed the umbilicals between the capsule and the instrument section, which fell away. Gagarin's re-entry then stabilized and he was ejected automatically as planned, and landed in a plowed field, watched by a bemused woman and child.

Longhaul space flights

According to Féderation Aeronautique International rules, a manned spaceflight was to be one in which a passenger flew in a craft as it was launched and as it landed, which is what the Soviets claimed. The fact that Gagarin had landed by parachute separately from his craft—like all Vostok pilots after him—was not revealed until much later.

The U.S. had launched two Mercury astronauts on sub-orbital flights in May and July, when on August 6, 25 year-old Soviet cosmonaut Gherman Titov—still today the youngest space traveler—was launched aboard Vostok 2. Titov spent a day in space, because he had to land in the prime recovery zone in daytime—in the Soviet Union—which would only be reached after 17 orbits, because of the Earth's rotation. It was revealed later that Titov was spacesick but did manage to sleep for a while and to eat and drink using tubes. The spacesickness was caused mainly by the effect of weightlessness disturbing the sensitive inner ear balancing mechanism, so the effect is similar to Earth-bound travel sickness.

Titov was feted as a hero and the U.S. prepared for its own, belated orbital flight, which took off in February 1962 carrying the legendary John Glenn, who was followed by Scott Carpenter—both making three-orbit flights—in May 1962.

In August 1962, the Soviet Union performed what initially appeared to be a spectacular feat—two Soviet cosmonauts "meeting" in space. Vostok's Andrian Nikolyev was launched first on August 11 and a day later, Vostok 4 was launched crewed by Pavel Popovich. The two craft passed within three miles (5 km) of one another as their orbits coincided. However, this was not a meeting, neither was it a true rendezvous by two maneuverable craft changing their orbits, but the Soviet government and gullible western media perpetuated this apparent Soviet lead in the Space Race. Despite these untruths, Nikolyev extended the manned spaceflight record to almost four days. America seemed to be falling badly behind.

The Vostok curtain-call came in June 1963 and not surprisingly, featured yet another spectacular. Valentina Tereshkova became the first woman in space on June 16, 1963 when she was launched on Vostok 6, and made a brief co-incidental rendezvous a few miles from Vostok 5, manned by Valeri Bykovsky who had been launched two days earlier. Tereshkova became the latest Soviet hero.

A former cotton mill worker and amateur parachutist, Tereshkova had a very unhappy time, feeling very sick most of the time, and even pleaded to come down earlier than planned.

Bykovsky on the other hand made what was to be the longest solo manned space flight in history, lasting about five days. It is a solo record that is likely to last for a very long time.

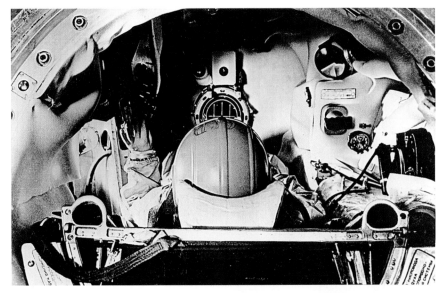

Left: The inside of the Vostok capsule reveals a more spacious cabin than the U.S. Mercury spacecraft.

Below: The first woman in space, Valentina Tereshkova, is pictured at the end her unhappy mission in June 1963.

The Mercury Project

In September 1958, the 43 year-old National Advisory Committee for Aeronautics (NACA) become known as the National Aeronautics and Space Administration (NASA). Its first act was to select a design for America's first manned spacecraft and a name for the project. NASA chose Mercury.

Above: A depiction of the high-risk re-entry of the Mercury capsule as it enters the Earth's atmosphere, protected by its heatshield.

A year later, the first Mercury pilots were selected from hundreds of military servicemen, many of whom were never even tested. The Mercury seven astronauts became household names even before they flew. They were viewed like knights in shining armor, setting out to fight the Cold War—in the battleground of space.

Due to the limited power of U.S. launchers, their capsule was so small that the astronauts joked that they did not get into it, they put it on. Despite this, Mercury was more sophisticated technologically than Vostok. Mercury was 9 ft 5 in (2.76 m) high with a maximum diameter across the base of 6 ft 1 in (1.85 m). It weighed about 2,980 lb (1,351 kg) at launch. The base was covered with an ablative heatshield, and the capsule had to be carefully orientated during re-entry so the

heatshield was pointing in the right direction and making the right angle of entry into the atmosphere.

The solid propellant retro rocket pack was attached to the heatshield and was normally deployed after firing and before re-entry. More than a hundred displays and consoles were mounted in front of the astronaut's face, providing spacecraft orientation, navigation, environmental, and communications control. The craft was equipped with a periscope and had a rectangular window on all but the first manned flight, which had a circular porthole.

Maneuverable spacecraft

The attitude of the Mercury capsule could be altered and controlled by the use of an airplane-like joystick commanding the release of short bursts of hydrogen peroxide gas from ten control thrusters located on various parts of the craft. These movements could be managed by the automatic stabilization and control system (ASCS), the rate stabilization and control system (RSCS), or by manual proportional control—fly-by-wire, a combined manual/electrical system.

The Mercury launch escape system comprised of a cluster of solid propellant rockets on a tower on top of the capsule. For egress, later models had an entry-exit hatch that could be exploded in the event of an urgent evacuation.

Mercury descended into the sea under a drogue and main parachute, and deployed a landing bag between the heatshield and the main body of the craft at the rear, to cushion the splashdown, which would otherwise have caused a 10G jolt.

Before the first manned mission, various Mercury systems and procedures, such as aborts, launch escape systems, heatshield, aerodynamics, and telemetry were tested during a series of test flights starting in 1959, featuring Little Joe, Atlas, and Redstone vehicles and the primates, Sam, Miss Sam, Ham, and Enos.

The Little Joe was a hybrid rocket with the performance of the Redstone but which cost five times less. It comprised of Castor or Pollux and Recruit solid rocket motors and was launched from Wallops Island, Virginia.

These Little Joe, Redstone, and Atlas tests were

not all successes. One of the worst moments—
although most comical—was the aborted lift-off of
a Redstone in November 1960, which culminated
in the accidental firing of the escape tower (without
the capsule) and then the deployment of the
capsule's landing parachutes, all while still sitting
on the launch pad!

Then there was Mercury Atlas 3—aiming for
the first Mercury orbital flight—which exploded
during launch, just 12 days after Gagarin's triumph.
This was another low point for the U.S.'s prestige
in the grim days of the early 1960s.

Left: Ham, the
chimpanzee who flew on
a sub-orbital Redstone-
boosted Mercury mission,
paving the way for
America's first man in
space, Alan Shepard.

Below: Little Joe
boosters were used for
several tests in the
Mercury program,
including the operation of
the launch escape system
pictured here.

America's First Astronauts

U.S. Navy commander Alan B. Shepard could have been the first man in space. However, after a troublesome Mercury Redstone 2 flight with the chimp Ham, Wernher von Braun insisted that another unmanned Mercury Redstone test would be needed. This flew on March 24, 1961, although the flight was only sub-orbital. It could have carried Shepard. As events unfolded, Gagarin was launched 19 days later, and the honor quite rightfully went to him.

Right: The historic lift-off of Alan Shepard from Pad 5 at Cape Canaveral on his sub-orbital flight aboard *Freedom 7*.

Gagarin had orbited the Earth rather than performing merely a brief sub-orbital mission, which was the objective of the Mercury Redstone 3 mission. The manned mission was eventually launched from Pad 5 at Cape Canaveral, with astronaut Shepard aboard his capsule named *Freedom 7*, on May 5, 1961.

"I have lift-off, the clock is started," he reported as if to announce the start of America's new era. The first American in space, Shepard made a flawless 15 minute 28 second flight on a suborbital trajectory. "Oh what a beautiful view," he said at apogee, probably trying to convince himself since he was looking out of a periscope and had forgotten to remove the black and white filter. Shepard was able to manually orientate his spacecraft during the brief spaceflight and all systems worked well. He splashed down safely and was recovered by a helicopter as planned.

Unlike the Soviet launches, Shepard's was covered live on TV. The open nature of America's space program was a credit to the nation. Rightfully, Shepard was treated as a national hero and awarded an honor by President Kennedy. The flight may have been just 15 minutes, but America was on its way at last.

Buoyed by the reaction to Shepard's success, President Kennedy became convinced that space was the arena in which the Cold War was to be fought, and prepared to make a dramatic announcement on May 25.

Six more sub-orbital missions were planned to give each Mercury astronaut experience before pressing on for orbital flights, but sensibly this was reduced to three, with Gus Grissom going next, to be followed by John Glenn.

Grissom was launched on *Liberty Bell 7* on July 21, 1961, and his mission was also a success, lasting 15 minutes 37 seconds. Shepard had called his spacecraft *Freedom 7* because its production number was seven, but Grissom named his capsule with the "7" representing the Mercury astronaut team, a custom that was followed by later members of the team when their turn came to fly.

Almost a tragedy

Up to splashdown everything with the *Liberty Bell* mission went well. Then the trouble started. As Grissom was preparing to exit the hatch for recovery by helicopter, the first "explosive" hatch blew accidentally. Water poured in. Grissom jumped out and almost drowned because he had forgotten to seal the hose connection of his suit, and water leaked though the neck ring dam. He was also weighed down by souvenirs from his craft that he was going to give to friends and colleagues.

Flailing in the water, Grissom was angered to see a photographer taking pictures of him from one of the recovery helicopters. His hand gestures pleading for aid were seen by the pilots as an indication that all was well. When he was finally snared by a helicopter wire, Grissom was hauled 30 ft (10 m) underwater before being winched up from the sea. Meanwhile, fearing that the waterlogged capsule would be too heavy, another helicopter pilot let go of the craft and Grissom's capsule dropped to the bottom of the Atlantic. It was a sad end and although Grissom was not officially blamed for the incident—though it had been suggested he accidentally hit the arming device—the stigma lived with him.

Liberty Bell 7 was raised to the surface 40 years later but Grissom was not alive to witness the event or the recovery of evidence from the remains of the capsule that appeared to vindicate him. After this mission, the third sub-orbital Redstone mission was canceled and Glenn moved from sure anonymity to fame as he was assigned the first fully orbital mission.

Mercury Orbital Flights

Before the first Mercury manned orbital flight, two test missions were flown, with a robot "astronaut" flying a successful one-orbit mission in September 1961, and the chimpanzee, Enos, a two-orbit mission—reduced from three by technical problems—in November 1961. It was now time for John Glenn to make his historic flight.

A merica's answer to Titov's 17 orbits was to be a three-orbit affair. John Glenn, the freckle-faced, ginger-haired pilot with a boy scout image had been assigned to the most prestigious of all the Mercury missions because the third Redstone sub-orbital flight had been canceled. From flying what would have been one of the most anonymous of missions, Glenn was catapulted to fame.

The mission of *Friendship 7*, as Glenn named his capsule, was delayed many times in late 1961 and early 1962. America was getting impatient. Finally, on February 20, 1962, Glenn rode the tetchy Atlas ICBM into the skies, and on reaching orbit exclaimed "the view is tremendous."

Listening live to the mission, America was in awe. At last, the country was catching up with the Soviet Union. Glenn assessed the maneuverablity of the craft in several modes, operating the instruments automatically or manually, in pitch, yaw, and roll. His mission

Right: Gordon Cooper's Atlas D booster rises from Pad 14 at Cape Canaveral on May 15, 1963, on the final Mercury flight.

became even more famous when flight controllers suspected that the craft's heatshield might be loose. Fortunately, the scare was a false alarm discovered only after minutes of hand-wringing, prayers, and worry as Glenn plunged through the atmosphere during the high temperature re-entry in which communications were impossible.

He emerged into radio contact, exclaiming that the re-entry had been a "real fireball, boy, I had bits of that retropack coming off all the way through." The triumphant Glenn splashed down in the Atlantic Ocean after the five-hour mission and America breathed a sigh of relief.

Later, Glenn was given a New York tickertape parade and seemed destined for greater things. He never flew in Gemini or Apollo, but at the age of 77, and by now a veteran politician who had once run for the Presidency, he again captured America's hearts with a flight on the Space Shuttle in 1998 (*see page 92*).

The problems stack up

Glenn's fortunes contrasted with those of the next Mercury astronaut, Scott Carpenter, who was to fly a repeat three-orbit flight launched on May 24, 1962. He named his capsule *Aurora 7*. Carpenter's flight plan was busy and his desire to enjoy his experience of exploring the unknown and to perform science led him to make mistakes which resulted in loss of attitude control propellants, and nearly ruined retrofire and re-entry. Carpenter overshot his landing by 250 miles (400 km), was called "careless" by NASA, and never flew again.

This was in some ways unfair, given the crammed flight plan, and it was rather ironic that the next pilot, Wally Schirra, was praised for flying a "textbook mission," which basically involved him sitting with his arms folded and drifting for long periods without using any maneuvering propellant. Carpenter got a raw deal.

Schirra's mission in *Sigma 7* lasted about six orbits on October 3, 1962, and preparations were made for what was to become the sixth and final manned Mercury mission by Gordon Cooper on May 15, 1963, which extended the U.S. flight time record to more than a day.

Cooper had fallen asleep in his capsule, *Faith 7*, the day before as he waited for a blast-off that was eventually canceled, but his casual style, magnified by a laid-back Oklahoma drawl, was deceptive. Cooper was a hot-shot pilot and flew an heroic

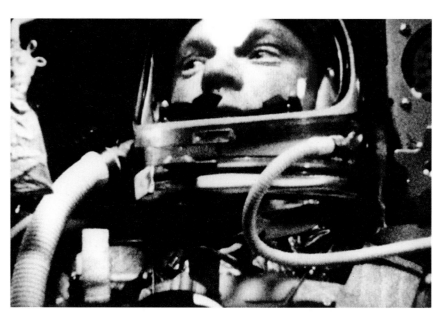

mission as various systems malfunctioned and eventually broke down. He gave a long list of malfunctions as the re-entry approached: "…things are beginning to stack up a little." He added, "other than that things are fine." Cooper flew a manual re-entry, landing almost on-target close to the U.S. Navy recovery ship the USS Kearsage, in the Pacific Ocean, after 22 orbits.

Alan Shepard's planned two-day Mercury flight was canceled. NASA moved on to Gemini—and the moon.

Above: An onboard picture of John Glenn during his epic three orbits in *Friendship 7*, which did so much to restore U.S. pride in February 1962.

Below: Wally Schirra's *Sigma 7* capsule is recovered after his nine-hour Mercury flight.

The X-15

In June 1952, NACA had recognized the need for research into manned spaceflight and directed its laboratories to study the problems involved in getting a man into space—and bringing him back. At this early stage in space research almost all the relevant technologies were as yet untested.

Two years later it was decided to establish a "spaceplane" program to develop and study the technologies required. The spaceplane was to fly at Mach 6 at 250,000 ft (76,000 m)—almost 50 miles (80 km) high. It would be air-launched like previous rocket planes but eventually a rocket-boosted ascent was anticipated.

On July 9, 1954, a Joint NACA-U.S. Air Force and U.S. Navy committee agreed the program and by November 1955, North American Aviation (NAA) were awarded the contract to build the spaceplane, and Reaction Motors the 57,000 lb (25,850 kg) thrust, throttleable XLR-11 engine. The X-15 was born.

Test pilots were selected from the three agencies, including a certain Neil Armstrong, while NAA's Scott Crossfield would make the first demonstration flights.

The first X-15 was completed in October 1958, the same month that NACA became NASA, the National Aeronautical and Space Administration. In June 1959, Crossfield made the first glide flight, and the first flight to Mach 2 in September.

Right: Joe Walker, the highest man in an aircraft—the X-15 rocket plane—died in a mid-air collision in 1966.

Main picture: The X-15 undergoes post-landing checks as the B-52 mother craft flies overhead in a traditional ceremony at the end of the mission.

Astounding height and speed

Gradually the X-15 was turned over to agency pilots, with speeds and altitudes of Mach 3 and 136,000 ft (41,450 m) being attained in 1960. A new XLR-99 engine was also introduced but almost killed Crossfield when it exploded during a captive test firing. Gradually the Mach number and altitudes went up, and by 1961 the X-15 had reached Mach 6.04 and 217,000 ft (66,150 m).

On July 17, 1962, U.S. Air Force pilot Robert White became an astronaut when he reached space—314,750 ft (95,950 m) or 59 miles (95k m). The X-15 altitude record was achieved by NASA's Joe Walker on August 22, 1963 when he reached 67 miles, his third "astro-flight."

By now, the X-15 was demonstrating and testing spaceflight technologies and carrying experiments to support NASA's Apollo program. One mission carried a

Saturn ablative test among its suite of experiments. Flights in 1967 featured two significant and different milestones. First Pete Knight flew a modified X-15 to a record speed of Mach 6.70 on October 3, but on November 11 Mike Adams was killed in a crash when the craft went into a spin. He had flown to an altitude of 266,000 ft (81,000 m) and become the 12th X-15 astronaut on this mission. The program ended on October 24, 1968 after 199 flights.

While the X-15 was flying, the U.S. Air Force

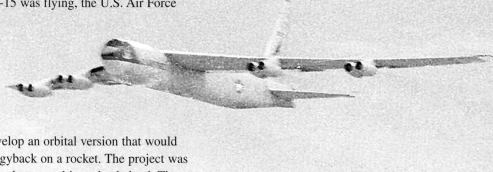

had plans to develop an orbital version that would be launched piggyback on a rocket. The project was called Blue Scout but was ultimately shelved. The Air Force also had plans to develop a new orbital spaceplane. Derived from X-15 technology, the vehicle was called X-20 Dyna Soar, and in 1962 six military astronauts were chosen to fly it. The two-crew Dyna Soar was to make glide flights in 1964, followed by two unmanned orbital flights launched on Titan 2 boosters, with the first manned flight set for 1966. There would be 12 flights every three months and a budget of $1 billion.

Conceived before Sputnik 1, Dyna Soar seemed the logical way to go—a reusable orbital manned spaceplane just like the 1950s depictions of the future of

space. But with the Cold War and Space Race—and excessive cost overruns, the X-20 was doomed and was canceled before it really got started, in 1963. Instead, the U.S. Air Force opted for a Manned Orbital Laboratory, MOL, based on NASA's second generation manned craft, Gemini. However, MOL was also canceled in 1966.

Below: The winged Dyna Soar manned spacecraft was to have been launched on the Titan II missile.

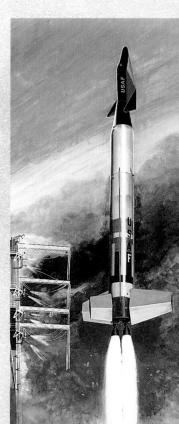

CHAPTER FOUR

RACE TO THE MOON

"We choose to go to the moon..."

The Apollo "Big Three,"
left to right: NASA's
Robert Seamans,
Wernher von Braun,
and President Kennedy.

When Sputnik I orbited the Earth, Americans could see its rocket stage fly overhead and hear the radio broadcasts of its "beeper." They were painfully aware that the Russians had beaten them to space. More importantly, Sputnik showed that the Soviets were also ahead in missile technology and had more powerful ICBMs. It was this military threat that would drive the United States to place men on the moon in an effort to demonstrate their technological lead.

The U.S. took the launch of Sputnik personally. The nation that had "won the war" and built the atomic bomb thought that its lead was unassailable. National pride was further damaged when Gagarin was launched.

Regaining the initiative was an "urgent national need," according to the title of the address given to Congress on May 25, 1961 by the relatively new and youthful President John F. Kennedy.

"It is time for a great new American enterprise—time for this nation to take a clearly leading role in space achievement, which in many ways may hold the key to our future on Earth," he stated.

Then he spoke the words that would launch the Apollo program: "First, I believe that this nation should commit itself to achieving the goal, before this decade is out, of landing a man on the moon and returning him safely to the Earth. No single space project in this period will be more impressive to mankind, or more important for the long-range exploration of space; and none will be so difficult or expensive to accomplish...

"I believe we possess all the resources and talents necessary. But the facts of the matter are that we never made the national decisions or marshaled the national resources required for such leadership. We never specified long-range goals on an urgent time schedule, or managed our resources and our time so as to ensure their fulfilment."

The Apollo "Big Three," left to right: NASA's Robert Seamans, Wernher von Braun, and President Kennedy.

Committing the nation

Kennedy inspired the American people to pursue a dream "impressive to mankind," and in the next eight years, the nation transformed itself. The pioneering astronauts were national heroes, their space missions were a national pastime. "It will not be one man going to the moon—if we make this judgment affirmatively, it will be an entire nation. For all of us must work to put him there," Kennedy had said to Congress.

Perhaps Kennedy's greatest speech on the nation's space effort was given on September 12, 1962, at Rice University: "But why, some say, the moon? Why choose this as our goal?… We choose to go to the moon... (applause)...We choose to go to the moon in this decade and do the other things, not because they are easy, but because they are hard, because that goal will serve to organize and measure the best of our energies and skills, because that challenge is the one that we are willing to accept, one we are unwilling to postpone, and one which we intend to win…".

The 1950s dreams of men on moon bases, huge space stations, and space shuttle taxis flying to and forth, men on Mars, and flying to the stars were put on hold. Mankind would go the moon as quickly as possible and as cheaply as possible, with not much thought what to do next. But the simple fact is that without a kick from the Cold War, it is unlikely that some of those dreams would have come true in the 20th century. NASA was directed to work out a way of getting to the moon and back by December 31, 1969—and to do it before the Russians. Kennedy's goal would be achieved.

Getting There

President Kennedy committed the United States to reaching the moon despite his country having achieved just five minutes of weightlessness in space. Alan Shepard's *Freedom 7* flight had been a milestone, but getting to the moon was in the big league.

It was to be the biggest technological challenge of the age, as building the pyramids had been for the ancient Egyptians. What's more, NASA was charged with the task of doing it in only nine years. Many potential plans of possible moon expeditions were already in the pipeline at NASA, and three came to the fore. The first and in a way the simplest was the Direct Ascent approach. Build a huge booster, send the spaceship to land directly on the moon, with enough rocket power to take off and fly right back to Earth. The booster would be the Nova, with a thrust of 40 million lb (18 million kg). However, the huge cost and the technological challenge was too great. After all, the U.S. had only just developed and flown the Atlas with a thrust of 367,000 lb (166,470 kg), and even the mighty Saturn 5 rocket would only develop 7.5 million lb (3.4 million kg) thrust.

The next proposal was equally logical: build the entire moon craft in Earth orbit by sending it up in pieces on smaller rockets. The spacecraft would then fly to and from the moon in a similar manner to the Direct Ascent approach. This was called the Earth Orbit Rendezvous plan. A rocket to do the job was already in the pipeline: the Saturn C-5, a sort of mini-Nova designed by Wernher von Braun. A useful spin-off from this method would be the creation of a space station in Earth orbit. This was already NASA's goal—a scientific base in space.

Multi-stage mission

The big problem was bringing the payload weights down to make the mission possible using Saturn C-5s—and keeping to the budget. NASA engineers came up with an alternative—use just one Saturn C-5 by flying a two-part spaceship to the moon. It was the cheapest option.

The spacecraft would be flown into Earth orbit with three crew and then fired off to the moon using the third stage engine of the Saturn C-5. The craft would fly into lunar orbit and a "lander" comprising descent and ascent stages would detach, flying two crewmen onto the surface, using the descent stage engine. The crew would later take off in the ascent stage, using the descent stage as launch pad, rejoining the main craft and its single

crewman in a rendezvous, and docking in lunar orbit.

The lander would be detached and the main craft would fire itself out of lunar orbit and head home, making a 24,000 mph (38,600 km/h), 4,000°F (2,200°C) plunge into the Earth's atmosphere, landing under three parachutes into the sea.

The engineers who developed this plan were led by John Houbolt of NASA's Langley Research Center. NASA chiefs at first thought he was crazy but gradually it dawned on them that the mission was possible and the only realistic way to go.

The Lunar Orbit Rendezvous mission, as it was called, was recommended and adopted in June 1962. It may have been the cheaper option but was also the most risky, with little room for error—but it was the one with which NASA felt it could meet Kennedy's extraordinary deadline.

The next decision was what to call the project. NASA had already named the first U.S. manned spacecraft Mercury after a figure in Greek mythology who was the messenger of the gods. The next manned spacecraft series, designed to carry two astronauts, was to be later named Gemini after a constellation with the main stars Castor and Pollux, named after characters in Greek mythology who rode horses across the sky. True to NASA's new moon mission, Apollo was more versatile. He was the Greek god of poetry, music, prophecy, and archery.

So, Apollo it was.

Above: John Houbolt, who developed the Lunar Orbit Rendezvous mission, which was eventually adopted for the Apollo program.

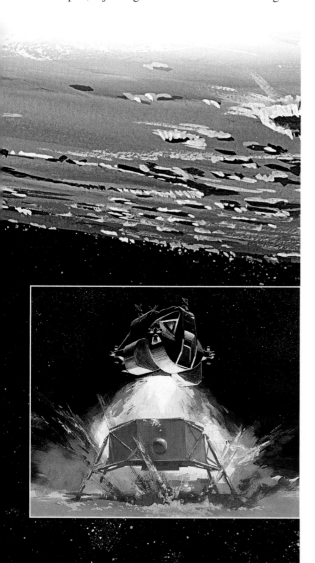

Left: The Apollo command and service modules, with the lunar module attached, enter lunar orbit with the firing of the Service Propulsion System engine, while **inset**, the lunar module ascent stage takes off from the moon, using the descent module as a "launch pad," at the end of the lunar exploration.

The Soviet Plan

The prestige of being the first nation to land a man on the moon was incalculable. While the U.S. announced its Apollo plans, the Soviet Union said nothing, other than that its goal was also the moon. Crucially, it would be another four years before the Soviets settled on a definite plan.

The Soviet lunar landing program was code-named L-3, signifying correctly that it was one of a number options. Full details of the program were not revealed until the 1980s, although American intelligence probably had an idea what was going on at the time.

If Apollo was an ambitious program, then L-3 was almost foolhardy and surely doomed to fail. L-3 depended on the Soviet mega booster, called the N-1, an extraordinary, almost uniformally tapered vehicle with a very long payload shroud at the top, in which the manned moon landing craft was placed.

The N-1 had a base diameter of about 50 ft (15 m) and was just over 300 ft (91.5 m) high. The first stage was equipped with 30 liquid oxygen-kerosene engines called NK-33, with a thrust of about nine million lbs (four million kg). These would operate for 110 seconds and would all have to work simultaneously, with equal thrust.

The second stage was powered by eight similar engines, called the NK-34, which were to operate for 130 seconds. The third and final stage of the booster itself was powered by similar engines called NK-39, which fired for 400 seconds.

If the giant stack actually worked, the Soviet moon ship would be in Earth orbit. The moon landing spacecraft consisted of four main units: two more rocket motors, a lunar lander, and a lunar orbiter, in which a two-person crew would fly to the moon.

The first rocket motor, called a Block G, would send the combination on a flight toward the moon and then be discarded. Next a Block D engine was to fire to place the moon craft into lunar orbit with an eventual low point of 10 miles (16 km).

Complex and risky

The lone lunar lander cosmonaut would put on a spacesuit and make a spacewalk to transfer into the lunar module below, leaving his companion in the mother ship, which would then separate from the Block D/lander combination. The Block D engine would fire for the descent burn to the surface, but just before impact at an altitude of about one and half miles above the surface, the 15 ft (4.57 m) high lander would separate and make a soft landing using its own engine.

The Block D stage would crash land.

The lone cosmonaut would exit via a circular hatch and down a ladder onto the lunar surface, planting the Soviet flag. The first man on the moon? The pioneer would stay just an hour before returning to the cabin with rocks and samples. The entire lunar module would take off from the moon using the same descent engine.

The lunar module would dock with the lunar orbiter and the moonwalker would transfer back during an EVA. The lunar lander would be jettisoned and the lunar orbiter, based largely on a new Soyuz manned spacecraft, would fire its engine and head for home, making a high speed re-entry into the Earth's atmosphere.

These plans depended on successful early testing of the Soyuz spacecraft, in Earth orbit—and later the N-1 booster. The Soviet Union also developed a manned moon program called the L-1, which would carry two cosmonauts once around the moon on a "lunar-loop" and back to Earth. This would provide a prestige-boosting prelude to the lunar landing.

Left: The giant N-1 booster, designed for the Soviet attempt to land a man on the moon, was equipped with 30 first stage engines.

Right: The entire one-man lunar lander would take off from the moon using the same engine that had performed the landing.

Below: The "Block D" spacecraft, based on the Soyuz, would fly to the moon and back.

The Voskhod Gamble

Space coups, such as Valentina Tereshkova's flight, were having an effect on American morale. Nikita Khruschev, very much the front man for the Soviet space program, pushed for more. In 1964, he ordered that three cosmonauts be launched into space to beat the U.S.'s Gemini spacecraft, which was to carry two crew.

Right, above: The Voskhod 2 spacecraft, with the extendable airlock used for the world's first spacewalk.

Right, below: Alexei Leonov during his EVA on March 18, 1965.

Below: The Voskhod 1 crew after landing at the end of their risky 24-hour mission, left to right: Boris Yegerov, Konstantin Feoktistov, and Vladimir Komarov.

Having three Soviets orbiting over the U.S. would be highly effective propaganda. There was, however, a big problem. The Soviets did not have a three-man spacecraft, so a one-man Vostok was converted in order to cram three crewmen inside. An unmanned test flight was made on October 6, and on October 12, the riskiest manned spaceflight in history began.

The weight of the vehicle increased to 11,900 lbs (5,400 kg). A back-up retro rocket was needed because launches on the upgraded Vostok launcher, the SL-4 with a new upper stage, took the craft into a higher orbit that would not permit a natural decay within the ten day deadline set for Vostok. The craft therefore had a solid propellant retro rocket added to the top of the spherical flight cabin.

Voskhod had most of its stuffing removed so that it could carry up to three crew lying side-by-side. This meant that the crew had no ejection seats—and no means to escape a launch failure. They also flew in tracksuits rather than spacesuits, and would be killed if there was a depressurization in space.

With no Vostok-like ejection capability and having to land inside the capsule, the spacecraft had to be equipped with a soft-landing retro rocket system. This fired just before touchdown, reducing the landing velocity to about 0.2 m per second.

After the unmanned test flight six days earlier, on October 12, 1964, cosmonauts Vladimir Komarov—the only actual pilot cosmonaut—a doctor, Boris Yegerov, and a space designer who had helped to reconfigure Vostok, Konstantin Feoktistov, were launched successfully.

They did little during their mission because they couldn't move much, and came home after a flight of just one day, as planned. The gamble had paid off and coincided with the apparently unconnected removal of Nikita Khruschev from power. However, Soviet propaganda went to work with superlatives about Russia's new multi-seat spaceship. The West lapped it up and got worried. The Soviets were apparently still leading the Space Race.

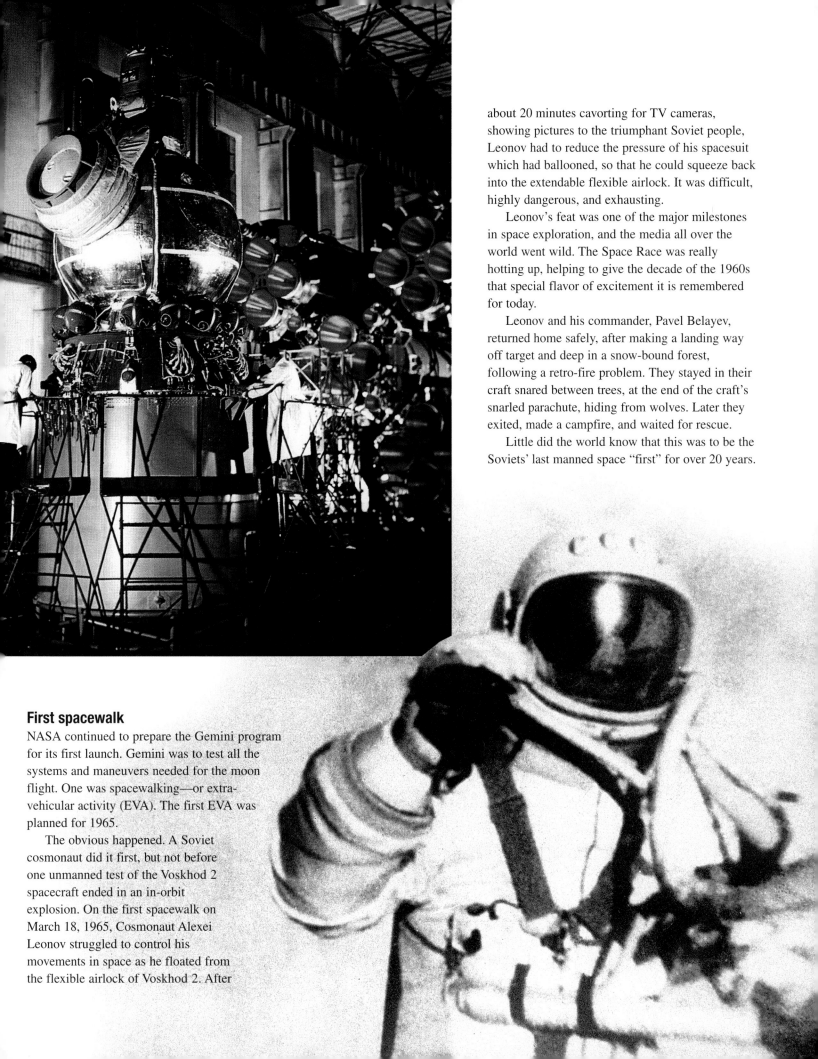

about 20 minutes cavorting for TV cameras, showing pictures to the triumphant Soviet people, Leonov had to reduce the pressure of his spacesuit which had ballooned, so that he could squeeze back into the extendable flexible airlock. It was difficult, highly dangerous, and exhausting.

Leonov's feat was one of the major milestones in space exploration, and the media all over the world went wild. The Space Race was really hotting up, helping to give the decade of the 1960s that special flavor of excitement it is remembered for today.

Leonov and his commander, Pavel Belayev, returned home safely, after making a landing way off target and deep in a snow-bound forest, following a retro-fire problem. They stayed in their craft snared between trees, at the end of the craft's snarled parachute, hiding from wolves. Later they exited, made a campfire, and waited for rescue.

Little did the world know that this was to be the Soviets' last manned space "first" for over 20 years.

First spacewalk

NASA continued to prepare the Gemini program for its first launch. Gemini was to test all the systems and maneuvers needed for the moon flight. One was spacewalking—or extra-vehicular activity (EVA). The first EVA was planned for 1965.

The obvious happened. A Soviet cosmonaut did it first, but not before one unmanned test of the Voskhod 2 spacecraft ended in an in-orbit explosion. On the first spacewalk on March 18, 1965, Cosmonaut Alexei Leonov struggled to control his movements in space as he floated from the flexible airlock of Voskhod 2. After

First Steps for Gemini

In order to design the Apollo system, NASA needed a series of spacecraft to test all systems in Earth orbit and to change orbit, perform rendezvous and docking in space, operate a flight computer, fly missions longer than those required for Apollo, and perform spacewalks. A larger version of the Mercury spacecraft was called into service, and again following Greek mythology, was logically called Gemini, since it would carry two astronauts.

Above: The design of the Gemini spacecraft was based on Mercury. The new craft had additional space and modules for rocket propulsion and environmental systems.

Gemini was to be launched on the U.S.'s second generation ICBM, the Titan 2, which made its maiden flight on March 16, 1962. The bell-shaped Gemini spacecraft came in two sections, a black re-entry-crew module and a jettisonable white-colored, two-part adaptor section. This housed control thrusters for orbital attitude and maneuvering system, four retro rockets, and a section carrying environmental supplies, such as oxygen and batteries for power. This section was jettisoned to expose the inner section for the retro burn.

After retro fire the inner adaptor section was also jettisoned to prepare for the capsule's re-entry. The complete spacecraft weighed about 8,000 lb (3,150 kg) and measured 18 ft (5.5 m) long and 10 ft (3 m) in diameter at the base of the white colored adaptor section.

The crew module was 11 ft (3.3 m) long and 7.5 ft (2.3 m) in diameter at the base. Each pilot had a small window, and the crew lay in ejection seats for launch escape capability. An entry hatch easy enough to open manually was included to allow a crewman to exit for EVA. The cockpit displays were similar to Mercury except there was no periscope. The spacecraft was the first to fly using a computer.

The craft was equipped with thrusters for the re-entry control system, which stabilized the spacecraft for the retro rocket burn. The nose section also contained the parachute system. The parachute opened on a lanyard which pulled the Gemini into a horizontal position for landing, rather than base-first as was the case in Mercury. This gave the first crew a shock, since they didn't expect such a violent snatch when the chute opened. Future crews knew what to expect.

New technologies

Some Gemini spacecraft carried rendezvous radar and associated equipment, and apart from Gemini 3, 4, and 6, on manned flights they carried unique oxygen-hydrogen fuel cells in the adaptor section, to generate electricity. A fuel cell is a device that changes chemical energy into electrical energy by the reaction between two chemicals, in this case liquid oxygen and hydrogen. A bi-product of the reaction between oxygen and hydrogen is drinkable water, which was used by the astronauts.

Two unmanned missions were flown in 1964 before the first manned flight in March 1965. The Titan II rocket used for the first Gemini was static tested at the temporarily renamed Cape Kennedy in January 1964, with the first manned Gemini 3 originally scheduled for October that year.

Gemini Titan 1 was launched from Pad 19 on April 8, 1964. The Gemini model stayed attached to the Titan's second stage in orbit as planned, and later re-entered the Earth's atmosphere. Gemini Titan II was to feature an operational model of the spacecraft. However, hurricanes Cleo, Dora, and

Ethel and technical problems delayed the launch in late 1964, and on December 8 the Titan II ignited but was shut down automatically on the pad after an engine experienced a hydraulics problem.

On January 19, 1965, the mission began, sending the Gemini craft into a deliberate sub-orbital path so that during re-entry, the heatshield would be tested to its limit. The 18-minute flight also successfully demonstrated the landing system and ocean recovery. All was ready for Gemini 3 and the next steps to the moon.

Left: The innovative fuel cell system that flew on seven manned Gemini missions consisted of tanks of liquid oxygen and liquid hydrogen.

Below: The Gemini Titan combination on Pad 19 at Cape Canaveral, which featured a gantry that was lowered rather than rolled back in the traditional fashion.

The Heavenly Twins

The manned Gemini program kicked off on March 23, 1965, with the launch of Gemini 3. The mission was a modest three-orbit affair, but with a "space first": the first manned maneuvers in orbit, a crucial test for Apollo. Flying with rookie pilot John Young was command pilot Gus Grissom, who had flown the second sub-orbital Mercury mission in 1961 and who was thus becoming the first person to make two spaceflights.

The remarkable Gemini program then soared ahead with nine more manned flights, ending in November 1966 and meeting all its goals during one of the most frenetic and exciting periods of the moon race. Missions were followed avidly by the press—and on international TV, thanks to the communications revolution, which resulted from the first use of geostationary orbiting satellites, such as Early Bird.

America's first EVA was completed by Edward White, the pilot of Gemini 4, on June 3, 1965, during a flight with commander Jim McDivitt lasting four days. The images of White's spacewalk, taken by McDivitt, remain space classics to this day. The mission was the turning point in the rivalry with Russia. After this, the U.S. never looked back.

A record eight days was accomplished by

Gemini 5, which was launched on August 21. Gemini 5 was the first Gemini to carry the electricity-generating fuel cells, which unfortunately malfunctioned, almost aborting the mission of Gordon Cooper and Pete Conrad, who had to power down and spent a rather boring spaceflight.

Gemini 7 was launched on December 4 on a 14-day mission. It acted as the rendezvous target for Gemini 6 on December 16. Gemini 6 had been thwarted by the loss of its original target, an unmanned Agena stage in October, and later by a launch pad abort on December 12. The two Gemini craft flew in formation as close as one foot (30 cm) from each other in the greatest spaceflight since Gagarin. The space foursome who achieved the feat were Frank Borman and James Lovell in Gemini 7 and Wally Schirra (a Mercury veteran) and Tom Stafford.

Simulating Apollo procedures

Astronauts Neil Armstrong and David Scott completed the first space docking on March 16, 1966, when Gemini 8 joined up with an unmanned Agena target rocket, simulating the ascent of a lunar module from the moon docking with a mother ship in lunar orbit. The historic achievement was marred when a thruster on Gemini 8 short circuited and could not be turned off, spinning the craft like a catherine wheel. Armstrong and Scott were close to losing consciousness but managed to make an emergency landing.

Gemini 9 was launched on June 3, and its pilot, Gene Cernan, completed a spacewalk lasting over two hours, which only illustrated how difficult it was to work outside a spacecraft without some form of restraint to compensate for the tendency to float upward. The commander was Tom Stafford.

On July 18, Gemini 10 used its Agena target vehicle's engine to boost its orbit to a record 474 miles (763 km) during a mission featuring John Young and Mike Collins, which also included a rendezvous and spacewalk over to the Agena 8. Space buddies Pete Conrad and Dick Gordon then flew Gemini 11 to an altitude of 850 miles (1,368 km) over Australia using its Agena 11 target rocket after launch on September 12.

The remarkably successful program ended with the landing of Gemini 12. Launched on November 11, it was crewed by James Lovell and Buzz Aldrin, whose spacewalk eased concerns about EVA work.

NASA felt confident that the major requirements for a moon mission had been fulfilled. The Soviets had not launched a cosmonaut since March 1965. Apollo beckoned.

The Lunar Pathfinders

The moon continued to be an elusive target after the first phase of exploration, led by the Soviet Lunas 2 and 3 in 1959, and it was not until July 1964 that there was another success.

U.S. Rangers 3–6 all failed in 1962–64 in attempts to land small capsules on the surface or to take high-resolution images of the moon before crash landing. Rangers 1 and 2 were Earth orbit tests. The next major triumph was to be Ranger 7's mission.

The 800 lb (314 kg) spacecraft plummeted into the moon's Sea of Clouds at 5,822 mph (9,316 km/h) on July 31, 1964 after transmitting 4,316 pictures, the final one showing a mottled surface with hundreds of small craters, with 1,000 times better resolution than anything seen by a telescope on the Earth.

Ranger craft were based on a spacecraft bus similar to that which was used for Mariner 2 *(see page 122)*, and mounted on it was a 4.9 ft (1.5 m) tapered payload body housing the 381 lb (173 kg) TV system, comprising six cameras. The images recorded on vidicons were scanned for TV transmission as they were taken.

Ranger 7 was followed by the equally successful numbers 8 and 9 in the series, which showed details of the Sea of Tranquillity and the crater Alphonsus in February and March 1965. These gave confidence that Apollo astronauts would be able to land but also concern about the number of craters and rocks. The next stage for NASA was to soft-land surveyor vehicles on the moon and to continue the reconnaissance of the moon from orbit, using Lunar Orbiters.

Meanwhile, the Soviet Union was also planning to explore the lunar surface and to orbit the moon with a series of Luna flights, which started inauspiciously in February 1963 and continued with no less than 12 more failures.

Pictures from the moon's surface

At last, the first soft-landing on the moon was achieved by the Soviet Union's Luna 9 on February 3, 1966, although it was not strictly speaking a "soft" one. The 220 lb (100 kg) surface capsule of Luna 9 traveled to the moon attached to the main spacecraft bus, and was equipped with a 3.3 lb (1.5 kg) TV camera. As the spacecraft plummeted to the surface a retro rocket was fired to slow the descent, and at 16 ft (5 m) above the surface the 1ft 10in diameter (55 cm) capsule was ejected, and like the main craft impacted on the moon at 13.7 mph (22 km/h).

The capsule came to rest and started transmitting, opening four "petals" which exposed the TV camera. This transmitted rather like a fax machine and rotated 360° to produce a 6,000-line panorama in 100 minutes, which could only stretch 0.9 miles (1.5 km) into the distance. The powdery surface was covered with small stones of various sizes, proving that the lunar dust—at least in the Ocean of Storms—was not deep.

In April 1966, another milestone was reached when Luna 10 became the first moon orbiter. It was

Right: As Ranger 7 plummeted toward the Sea of Clouds at a speed of 5,822mph on July 31, 1964, its suite of cameras took images with a resolution 1,000 times better than any telescope. The circle in the last frame shot indicates the point of impact.

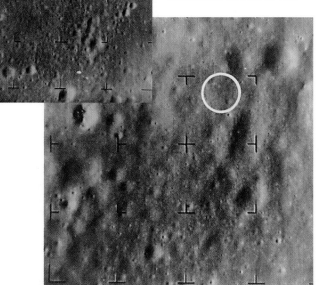

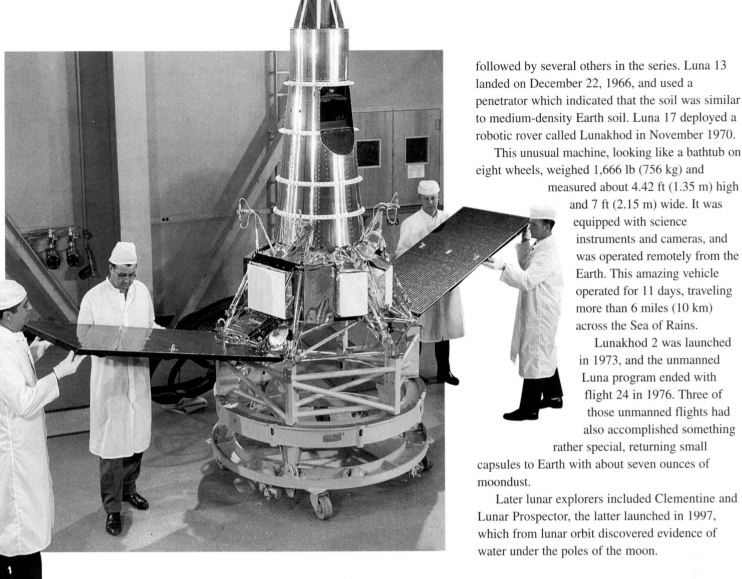

followed by several others in the series. Luna 13 landed on December 22, 1966, and used a penetrator which indicated that the soil was similar to medium-density Earth soil. Luna 17 deployed a robotic rover called Lunakhod in November 1970.

This unusual machine, looking like a bathtub on eight wheels, weighed 1,666 lb (756 kg) and measured about 4.42 ft (1.35 m) high and 7 ft (2.15 m) wide. It was equipped with science instruments and cameras, and was operated remotely from the Earth. This amazing vehicle operated for 11 days, traveling more than 6 miles (10 km) across the Sea of Rains.

Lunakhod 2 was launched in 1973, and the unmanned Luna program ended with flight 24 in 1976. Three of those unmanned flights had also accomplished something rather special, returning small capsules to Earth with about seven ounces of moondust.

Later lunar explorers included Clementine and Lunar Prospector, the latter launched in 1997, which from lunar orbit discovered evidence of water under the poles of the moon.

Above: The Ranger 6–9 fleet was equipped with six TV cameras, which took images recorded onto vidicon, scanned and transmitted them to Earth.

Right: The first craft to "soft land" on the moon was the Soviet Luna 9 capsule, which did not have a retro rocket but was deployed from the main craft at a height of about 15 ft, landing at a speed of about 14 mph.

The Moon Scouts

Two projects essential to America's Apollo program were Surveyor and Lunar Orbiter, the scouts searching for suitable landing places for the first astronauts. Landing humans on the moon would be a risky adventure even with prior knowledge about the lunar surface.

Below: A mock-up of the Surveyor spacecraft on the Earth. Five craft made successful soft-landings, helping to pave the way for manned landings.

The Surveyor lander was a 9.8 ft (3 m) high insect-like craft with three legs, spanning 14 ft (4.27 m) and weighing about 2,200 lb (1,000 kg) at launch. It carried a large solid propellant landing motor and a mast which at its

top held a solar array providing 85 watts of power and a high gain communications antenna. The main payload was a 16 lb (7.3 kg) black and white TV camera. The third Surveyor was also equipped with a Mechanics Surface Sampler, an articulated robot arm with a small shovel-scoop.

With Luna 9 on the lunar surface and Luna 10 orbiting overhead, the stakes were high for NASA as Surveyor 1 was launched on May 30, 1966. It soft-landed in the Ocean of Storms on June 2, and the TV camera beamed images live to viewers on the Earth.

The success seemed to illustrate the excitement of the moon race felt by the public, particularly as during the same week, Gemini 9 was launched and the first engineering model of 's Saturn 5 was rolled out to the launch pad at the Kennedy Space Center.

Surveyor 2 failed but number 3—equipped with a soil shovel—landed at another location in the Ocean of Storms in April 1967, bouncing twice and leaving footprints for the camera to reveal later. The shovel dug trenches of soil that was very like fine-grained earth. Clearly, landing an Apollo craft in thick and dangerous dust was not a likely scenario, as had earlier been feared. Later, Surveyor 3 was to receive some visitors—the crew of Apollo 12 (*see page 70*).

Surveyor 4 failed but Surveyors 5 and 6 landed safely in other potential Apollo touchdown locations in 1967, while Surveyor 7 performed a spectacular curtain-call, landing in the highlands close to the crater Tycho in January 1968.

Searching for a landing site

Meanwhile, the Lunar Orbiter program was becoming one of NASA's most successful, with all five craft successfully performing spectacular reconnaissance missions. Their prime aim was to provide NASA with 9 ft (3 m) resolution images of potential landing sites between 45° east or west, and within 5° north or south of the equator.

The Lunar Orbiter was an 835 lb (380 kg) craft equipped with an Eastman Kodak photographic system which involved 70 mm high-definition black and white aerial film being processed, rather like a Polaroid camera today, into a negative image which was scanned by a light beam and transmitted to Earth for image reconstruction.

Lunar Orbiter 1 was launched on August 10, 1966 and on August 23 transmitted back the first image, Earthrise over the moon's horizon, which did not seem to create the impact that a similar picture taken two years later would. The second Lunar Orbiter sent back an image of the crater Copernicus taken from an altitude of about 27 miles (45 km), which was dubbed the "Picture of the Century" by NASA.

The images were certainly spectacular, particularly Lunar Orbiter 3's photographs of the far side, including the crater Tsiolkovsky, number 4's view of the Mare Orientale impact basin, which looked like a dartboard, and the crater Tycho by number 5, the final craft, launched in August 1967. But spectacular images were not the raison d'être for Lunar Orbiter, which covered most of the moon back and front in a variety of orbits.

By the end of the Lunar Orbiter program NASA had been able to select 20 potential landing sites on the moon for Apollo missions, which were later slimmed down to eight. One of these targets was Site 2, at 0.75°N/24.2°E, in the Sea of Tranquillity—landing site of Apollo 11.

Above: This image of the 20-mile diameter crater, Kepler, was taken from an altitude of 34 miles from Lunar Orbiter 3 in 1967.

Left: The panorama of the terrain close to the Tycho crater taken by the final Surveyor, No 7, which landed in January 1968. The crater in the foreground is about 9ft in diameter and the distant hills are about 8 miles away.

APOLLO

The Apollo System

The Gemini, Surveyor, and Lunar Orbiter pathfinders had done their job. Now it was time for the real thing—the Apollo Saturn 5 and the trip to the moon.

Right: An early version of the three-crew Command Module, the only part of the Apollo system to return to Earth.

Below: The extremely lightweight Lunar Excursion Module (LEM) was nicknamed the "tissue paper" spacecraft. It was designed to fly only in space.

The Apollo space system comprised of three major components: a command module (CM) in which three crew were launched and flew to and from the moon; the service module (SM), and the lunar module (LM), originally called the LEM, for lunar excursion module. The CM and SM were also known as the CSM. Acronyms became an important part of space jargon.

The Apollo craft was fitted with a launch escape system (LES), a solid propellant rocket that could be used in the first 100 seconds of flight in case the Saturn 5 booster malfunctioned, pulling the CM to make a parachute landing in the sea. Otherwise, the LES was jettisoned after 100 seconds.

En route to the moon, the combined CSM separated from the S4B third stage of the Saturn 5, turned around, and docked with the LM nestled inside the S4B, extracting it, in what was called the transposition and docking maneuver. The S4B was then jettisoned. The combined craft flew to the moon, with the crew able to transfer to and from the two habitable modules via a transfer tunnel, once the docking probe had been removed.

The 12,000 lb (5,448 kg) CM was about 12 ft (3.66 m) high and wide, providing 235 cu/ft (8.29 cu/m) of space for the crew. The CM served as the flight deck, bedroom, kitchen, washroom— and "toilet." Urine was vented into space, while the crew had storage bags for defecation.

The CM display console, which faced the crew from their stations on three reclining seats on the floor, measured about 7 ft (2.1 m) across, with switches and dials for all the systems, including the flight computer and maneuvering thrusters and main service propulsion engine. In a small section at the foot of the couches was the navigation bay. The cabin atmosphere (originally pure oxygen) was at 15 psi (1 bar). A heatshield at the base of the CM protected the crew from 1,600°C (3,000°F) temperatures during the plunge into the Earth's atmosphere at a speed of about 25,000 mph (40,250 km/h).

The 54,000 lb (24,520 kg) SM was about 25 ft (7.6 m) long, with a conical rocket motor nozzle at its end and a large communications antenna on a short deployable boom. The service propulsion system (SPS) engine, which was vital for the lunar orbit insertion and trans-Earth burns, had a thrust of 20,500 lbs (9,307 kg). In addition to fuel, the service module held tanks of liquid oxygen and hydrogen for fuel cell tanks to convert into electricity and water as a bi-product.

Little more than kitchen foil

The LM, looking like a gangly insect and nicknamed "the bug," was a two-stage vehicle, 23 ft (7 m) high and 31 ft (9.4 m) wide across its four spindly landing legs. It weighed about 34,000 lb (15,436 kg) but was very fragile on the Earth, made from aluminum alloy, with a thin layer of insulation. The weight was made up largely of the hypergolic fuels in the ascent and descent stage tanks. Weight was at a premium—the LM was called the "tissue paper" spacecraft.

It comprised the descent stage, which was unmanned and contained the vital engine to perform the lunar landing, and an ascent stage on top, which housed the flight deck for the two crew, the commander on the left and lunar module pilot (LMP) on the right. The LMP was basically the systems engineer. Each would exit, crawling on their hands and knees onto a small porch on top of the ladder on one of the LM's four legs.

The descent stage was used as a "launch pad," as the ascent motor flew the crew into lunar orbit for rendezvous and docking with the CSM, manned by the lone command module pilot. Prior to re-entry into the Earth's atmosphere, the CM jettisoned the SM.

Making the Monster

Wernher von Braun's original Saturn C-5 rocket became known simply as the Saturn 5 once the Apollo program was up and running, with the building of the giant rocket being masterminded at NASA's Marshall Space Flight Center, Huntsville, Alabama. The Saturn 5 was designed to be the workhorse not just of the Apollo program but also of subsequent programs, such as a U.S. space station and even potential expeditions to Mars.

Right: A Saturn 5 second stage is prepared for testing at NASA's Marshall Space Flight Center, Alabama.

The Saturn was massive: 363 ft (110.7 m) high with a maximum diameter of 33 ft (10 m), but Apollo did not just involve building the rocket but also the massive structure needed to launch it. This was located just north of Cape Canaveral, on Merritt Island, which eventually became known as the Kennedy Space Center. The KSC today is still dominated by the building in which the Saturn 5 rockets were vertically assembled. The Vehicle Assembly Building, or VAB, it is now used to assemble the Space Shuttle. The VAB is 525 ft (160 m) high and could fit four United Nations buildings inside.

The Saturns were erected on Mobile Launch Platforms inside the VAB, which then moved slowly out of VAB's huge doors, on a crawler-way to either of the two launch pads, 39A or 39B. A planned pad 39C was never built. The platforms were the actual "launch pads," and included 410 ft (125 m) high mobile service structure gantry towers, which had two elevators and four access platforms to prepare the rocket and spacecraft.

The entire structures, weighing over 17 million pounds, were rolled down the three-and-a-half-mile crawlerways on a crawler transporter, a double-tracked vehicle the size of a football field. Each of the cleats of the tracks or "wheels" of the crawler weighs over one ton. The system is used today for the Space Shuttle, which is launched from the same two pads. Adjoining the VAB was the Launch Control Center, with four firing rooms, each with 470 sets of consoles and monitors.

All the power of a pocket calculator

The Saturn's five first stage F-1 engines generated a thrust of 7.5 million pounds. Each engine gulped almost 30,000 lbs (13,620 kg) of propellant every second, with a total load of 346,400 gallons (1.5 million liters) of liquid oxygen and kerosene. The S-1C first stage carried the vehicle to an altitude of 38 miles (61 km) and a speed of 8,350 mph (13,443 km/h) in 160 seconds. This was then jettisoned and the SII second stage was ignited. This was powered by five J2 cryogenic engines which consumed liquid oxygen and liquid hydrogen. The SII worked for six minutes and 30 seconds, by which time the rocket was 114 miles (231 km) up and traveling at about 15,300 mph (24,633 km/h).

The third stage was the re-ignitable S4B, also powered by one cryogenic J2 engine. Depending on the flight computer, the engine burned for about 150 seconds, achieving a speed of 17,400 mph (28,014 km/h), enabling the combination to enter Earth parking orbit.

Later the S4B was restarted to set the Apollo on course for the moon, on what was called the translunar injection burn. It fired for about 300 seconds, increasing speed to 24,400 mph (39,284 km/h), or escape velocity, to allow the crew to escape the pull of the Earth's gravity. After the LM had been extracted from the nose of the S4B, the spent stage sailed into deep space. It either went into solar orbit or impacted on the moon.

In its day the Saturn 5 had a very sophisticated flight computer. Called the instrument unit, it was a ring-like structure mounted around the rocket which measured the booster's acceleration and attitude, and calculated what corrections were necessary, commanding the engines' burn time. It also measured the booster's telemetry, electrical supply, and thermal conditioning system. The unit was 3ft (0.9 m) high, 21 ft (6.4 m) in diameter, and weighed over 100 lbs (45.4 kg). Despite this, it had a computing power less than one of today's simplest pocket calculators.

Below: A Saturn 5 is rolled out of the Vertical Assembly Building (later known as the Vehicle Assembly Building) at the Kennedy Space Center. The Launch Control Center is the smaller building on the left. Note the spare Crawler Transporters.

"Fire in the Cockpit!"

By 1966 the Apollo program was making astonishing progress, and NASA was almost ready to launch the first manned flight on a Saturn 1B rocket. The mission would be called Apollo 1 and would include the famous Mercury veteran Gus Grissom in its crew. Tragically, it was not to be.
A terrible accident would occur during simulation, resulting in the deaths of the three Apollo 1 crew and the grounding of all manned Apollo flights.

Below: A Saturn 1 booster is launched from Cape Canaveral with a live first stage and dummy upper stages during the first test flight phase of the Saturn program.

The first flight of the Apollo program took place on October 27, 1961, when a Saturn 1 booster was launched from Pad 37 at Cape Canaveral. Only the first S-1 stage was active, and powered by the engines that would be used on the operational Saturn C-1, later to be known as the Saturn 1B.

Three more flights followed, two of which dispersed thousands of gallons of water housed in the upper stage into the upper atmosphere as part of "Project Highwater," carried out to give engineers an idea of what would happen if tons of liquid propellants were released into the upper atmosphere in the event of a launch explosion. In the case of the Saturns, the release caused the formation of a huge ice cloud several miles in diameter. These flights, which ended in 1963, were followed by the Saturn 1 series, starting in 1964 and carrying dummy Apollo spacecraft and involving the test of the Saturn 5's planned restartable S-IVB stage, flying as the second stage on the Saturn 1. Three of the six tests carried active satellites, Pegasus, based around the SIVB and equipped with huge micrometeoroid detection panels, the last of which was launched in 1965.

Two Saturn 1s were then flown in 1966 carrying operational Apollo spacecraft on Earth orbital and re-entry tests, and were followed by the first Saturn 1Bs, one of which, designated Apollo 5, carried the Apollo lunar module on an orbital test flight, in January 1968. Just before this, on November 9, 1967 the first Saturn 5 was launched as Apollo 4, from Pad 39A at the Kennedy Space Center. The tumultuous lift-off, which shook the ground and stunned nearby observers with the apocalyptic noise, was a great success.

Events were gaining momentum. The first manned Apollo mission, on a Saturn 1B, had been scheduled for 1966 but was delayed to February 1967. It was called Apollo 1. Mercury veteran Gus Grissom was commander, with the first American spacewalker, Ed White, acting as senior pilot, and rookie Roger Chaffee, pilot.

The Apollo spacecraft was built by North American Aviation (NAA), which came as a surprise after the achievements of the McDonnell company with the Mercury and Gemini successes. NAA's selection was a political one. The company was not performing well with Apollo. Grissom's spacecraft was misbehaving and glitches were threatening the launch date.

Tragedy puts Apollo on hold

The crew was inside Apollo 1 on Pad 34 at Cape Canaveral on January 27, 1967 for a powered up "countdown to zero" test of the spacecraft on top of the unfueled Saturn 1B. Several glitches had already delayed the test and some people were getting rather tetchy, including a frustrated and swearing Grissom.

The crew had been in the capsule, pressurized with 100% oxygen, for

over five and a half hours when at 6:31 pm a spark from a short circuit in a bundle of wires that ran to the left and just in front of Grissom's seat started what became an inferno in seconds.

One of the astronauts, probably Chaffee, reported, "Fire, I smell fire." Two seconds later White said, "fire in the cockpit." Even under ideal conditions, it would take 90 seconds to get the hatch open. Because the cabin had been filled with a pure oxygen atmosphere, the fire spread extremely rapidly and the astronauts had no chance to get the hatch open.

The large amount of flammable material in the cabin allowed the fire to spread quickly. As White desperately tried to get the ratches loose, possibly with the help of Grissom, the garbled transmission from the cockpit sounded like, "We're fighting a bad fire—let's get out. Open 'er up." The transmission ended with a cry of pain, perhaps from Chaffee who had stayed in his position, nine seconds after the fire started.

Television monitors showed flames spreading from the left to right side of the command module as the cabin ruptured under the intense pressure and caused a surge of flames and heat. The unconscious crew died from smoke inhalation and burns.

It was all over in 13 seconds but was NASA's darkest hour. Apollo and America were grounded for 21 months. Had they lost the moon?

Left: Astronauts (left to right) Gus Grissom, Ed White, and Roger Chaffee were killed in the Apollo 1 spacecraft fire on January 27, 1967.

Right: The first giant Saturn 5 booster makes a thunderous lift-off from Pad 39A at the Kennedy Space Center on November 9, 1967.

To the Moon at Christmas

Twenty months after the Apollo 1 tragedy, the phoenix rose from the ashes and America was again on course for the moon. Without the exposure of defects in the Apollo system and the improvements made after the fire, including a safer nitrogen-oxygen atmosphere and a quick-opening hatch, it is unlikely that America would have made it to the moon by 1969.

C learly, a lot was riding on the first manned Apollo mission, called Apollo 7, which was to be the shakedown flight that Apollo 1 was to have conducted. The crew, commander Wally Schirra—the Mercury and Gemini veteran heading to become the first person to make three spaceflights—and rookies Donn Eisele and Walt Cunningham felt the pressure, and their tetchy behavior revealed the tension.

Schirra at times acted like a prima donna, and Eisele and to a lesser degree Cunningham were also deemed arrogant and never flew again. However, the 12-day shakedown was a brilliant success, including station-keeping with the upper stage of their Saturn 1B launcher, rendezvous simulations using the SPS engine, and the first in-flight TV programs broadcast to the world, which were appreciated by an eager public and set the standard for future Apollo broadcasts.

After the safe landing, the next plan was to fly Apollo 8 on a full simulation of the lunar landing but in Earth orbit, mainly putting the lunar module through its paces, including rendezvous and docking, flying the LM solo, and firing its descent and ascent engines. However, NASA was getting worried about the Soviet N-1 super-booster and a potential Soviet moon landing, and more urgently about some unmanned Soviet Zond lunar fly-bys and the stated plan to send two cosmonauts on a lunar-looping flight.

NASA decided to make the biggest gamble in

Above: Walt Cunningham, the Apollo 7 pilot, pictured in space during the 12-day shakedown flight in October 1968.

Right: Apollo 8 made the first manned flight into lunar orbit in December 1968, and returned with spectacular images of the lunar surface and the most famous image of the Space Age: Earthrise.

space history—fly an Apollo mission, without the lunar module, not just around the moon but to fully orbit it. The Apollo 9 crew became Apollo 8, and made history. Their names: Frank Borman, James Lovell, and William Anders. Launched on December 21, 1968 aboard only the third Saturn 5 rocket, the crew became the first to achieve escape velocity from Earth's gravity, at 24,200 mph (38,960 km/h).

The mission captured the imagination of the world, particularly as a result of the friendly TV transmissions from the crew as they became the first people to see the Earth as a small disk, slowly receding from them as the moon got bigger. After a three-day flight, Apollo 8 disappeared around the far side of the moon and out of radio contact with the Earth—the crew became the most isolated humans in history. The world waited.

As Apollo 8 came around the moon and back into contact with Earth, the crew gave the first descriptions of what it was like to see the moon from close quarters. "It looks like plaster of Paris," said Lovell. TV signals were beamed to Earth showing the rough terrain passing rapidly by the windows.

Historic broadcast

On Christmas Eve, the crew ended a TV transmission with "the Apollo crew has a message for you...". First Anders, then Lovell, and finally Borman read the opening passages from *Genesis*, the first chapter of the Holy Bible. "In the beginning, God created the heavens and the Earth…". It was one of the most remarkable and memorable transmissions in history and is often replayed in TV documentaries. The best was to come, however. As Apollo 8 came over the horizon on one of the orbits, the Earth was seen rising above the moon. Borman immediately took a camera and took the first images in black and white. Anders quickly loaded another camera with color film and snapped one of the most awe-inspiring and memorable images in history—Earthrise.

Lovell described the Earth as "a grand oasis in the vastness of space." The irony is that Apollo 8 is remembered more for the "Earthrise" photograph than it is for the moon.

Fine-tuning

As moon fever was building up in 1969, NASA had two more vital missions to perform before committing itself to a landing on the moon. Apollo 9 was to make the original Apollo 8 mission, conducting an all-up test of the total spacecraft, including a simulated lunar landing and take-off—all in Earth orbit. Then Apollo 10 would repeat the exercise in lunar orbit.

These flights were more for the scientists than the general public, but absolutely vital. James McDivitt, David Scott, and rookie Rusty Schweickart rode Apollo 9 into Earth orbit aboard the fourth Saturn 5 on March 3, 1969. They named their command and service module (CSM) *Gumdrop*, and the lunar module (LM) *Spider*. This was mainly to aid communications between the spacecraft.

On reaching orbit, Scott separated the CSM from the S4B third stage, turned around, moved toward the stage and plucked out the LM, in the transposition and docking maneuver. A spacewalk was then planned, with Schweickart wearing the lunar EVA suit, moving from Spider and into Gumdrop. However, he had

become sick soon after entering orbit, one of many space travelers to suffer from what is now known as Space Adaptation Syndrome (SAS), due mainly to the effect of zero G on the inner ear mechanism. It is suspected that Apollo 8's Borman had also suffered from SAS. When he felt a little better, Schweickart tested the spacesuit standing on Spider's porch instead.

Later, *Spider* separated from *Gumdrop* and became an individual spacecraft, and one incapable of returning to the Earth. The descent engine was fired to place it into a different orbit, then the ascent stage engine was fired, and the top half of the craft rendezvoused and docked with *Gumdrop*.

This highly successful ten-day mission was followed by Apollo 10's repeat performance which took two of its astronauts, Tom Stafford and Gene Cernan, to within nine miles (14 km) of the lunar surface. The third astronaut, John Young, became the most isolated person in history, flying solo around the far side of the moon, out of contact with the Earth. The mission, launched on May 18, 1969, was well covered on TV, and the broadcasts beamed by the astronauts were the most lighthearted yet, illustrated by the names of their spacecraft, the CSM *Charlie Brown* and the LM, *Snoopy*.

Right: The lunar module *Spider* is seen flying independently in Earth orbit during the Apollo 9 test flight in March 1969.

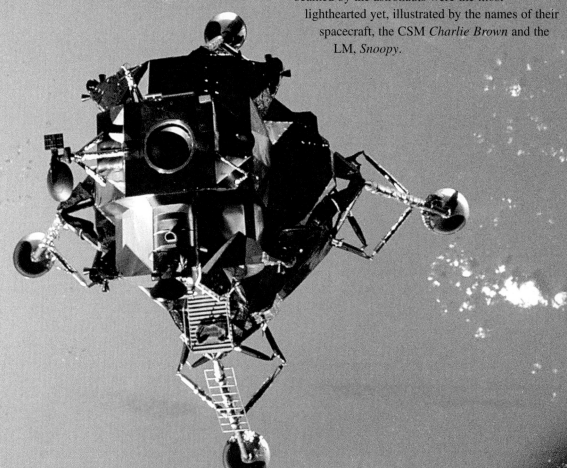

Entertaining the public

After entering lunar orbit, the riskiest mission thus far entered its critical moment. *Snoopy* separated from its mother ship, fired its engine, and swooped low toward the Sea of Tranquillity, the landing site selected for Apollo 11. "We is low, we're down among'em Charlie!" shouted an exuberant Cernan to Capcom Charlie Duke, alluding to the terrain, which included large craters and boulders. "Awe Charlie, we just saw Earthrise, it's got to be magnificent," Cernan shouted.

Later, as the descent stage separated and ascent stage engine fired, the euphoria gave way to sounds of impending disaster. A switch left in the wrong position resulted in the ascent stage gyrating wildly. "Son of a bitch! There's something wrong with the gyro," shouted an excitable Cernan. Control was restored. "There was a moment there, Tom..." said Cernan. He didn't need to continue. A special moment occurred when *Snoopy* docked with Young's *Charlie Brown*, and the crew headed home.

They set yet another new speed record during re-entry, as the command module plunged into the Earth's atmosphere at a speed of 24,791 mph (39,897 km/h), later splashing down in the Pacific Ocean near a recovery ship in the usual manner, after eight days in space. Apollo 11 was next on the launch pad—and this would be the big one.

"Tranquillity Base Here"

Eight years and two months after Kennedy's famous speech to Congress, the time had arrived. $25 billion had been spent, but the dream was about to become a reality. Unknown to the general public, the Soviet Union had already lost the race to the moon and America was merely racing itself, determined to land a man on the moon by the end of the decade as Kennedy had promised.

Six astronauts had already seen the moon from orbit, now two were to kick up the dust on it. Apollo 11 commander Neil Armstrong would be the first to set foot on the moon.

History is rarely as cut-and-dried as we imagine, however. Had the original Apollo 8 and 9 mission crews remained with these missions, Armstrong would have been back-up Apollo 9 commander and would have flown Apollo 12 instead. Apollo 11 would have been led by Pete Conrad, Apollo 8's back-up. Armstrong was just one brave and talented man in a team of equally talented individuals, any of whom could have been the first human being on the moon.

On the Apollo 11 mission, it was only natural that Armstrong would be the first out of the lunar module due to the layout of the cockpit and the way the hatch opened, but original checklists had the pilot going out first. The pilot in question was Buzz Aldrin, who did not take kindly to being bumped out of the history books.

Below: Neil Armstrong leads his team to the transfer van for the ride to Pad 39A for the Apollo 11 launch. The astronaut "walk-out" is described ominously by the press as the "last photo opportunity."

The world watches

Armstrong, Aldrin, and the command module pilot Mike Collins, were launched on July 16, 1969. Watched by thousands at Cape Kennedy and millions more over the world live on TV, the greatest mission began under the now familiar thunderous roar of a Saturn 5 from the Kennedy Space Center.

The flight to the moon followed the style of Apollo 10 and without much drama, as the mission was unfolded in detail by countless live TV programs. Everyone, it seemed, was flying the mission with this crew.

Entry into lunar orbit was achieved quietly and efficiently, reflecting the personalities of the crew and the seriousness of what was about to take place. On July 20, 1969, the lunar module *Eagle* undocked from the command module *Columbia*, fired its descent engine, and headed for a target site in the Sea of Tranquillity, which was thought to be relatively smooth.

The descent went well, with mainly Aldrin being heard giving instrument readings, but during the final approach a drama began to unfold. An overloaded computer protested with a series of alarms and it looked as though there might have to be an abort. Luckily, this type of malfunction had occurred during simulations, so it was decided to override the alarms.

The descent seemed endless and concern started to grow when Ground Control Capcom Charlie Duke shouted "60 seconds." This meant that Armstrong and Aldrin only had that much time to land or they would have to abort the descent.

The enigmatic and calm Armstrong was trying to find a safe place to land amid the craters and rocks. Aldrin told Armstrong there was 8% of fuel left in the tank. They needed some of that in case there was an abort. Sharing control with the automatic pilot, Armstrong flew Eagle over a crater which it was aiming toward, and flew further downrange. Armstrong told Aldrin that he had to get even further, to a smooth landing site. The fuel quantity gauge was now lit-up.

Aldrin could be heard giving instrument readings to Armstrong; "Down two and a half... forward... forward... two and a half... picking up some dust... faint shadow... drifting to the right a

little." Duke interjected, "30 seconds!" Time seemed to stand still. "Drifting right… contact light," continued Aldrin… "descent engine command override off. Engine alarm off. 413 is in." The craft landed, with just 20 seconds of fuel left in the tank.

After a moment's silence the calm voice of Armstrong announced, "Houston, Tranquillity Base here. The Eagle has landed." The breathless Duke didn't expect the formal name of the landing site and replied, "Roger, twan… Tranquillity, we copy you on the ground. You got a bunch of guys about to turn blue. We're breathing again. Thanks a lot."

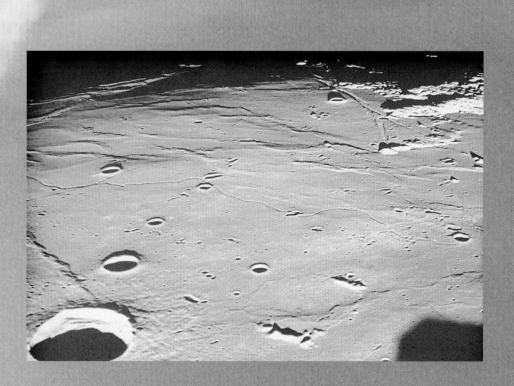

Right: Condensation billows around the midsection of the Saturn 5 as Apollo 11 begins its journey to destiny, while **inset** the region of the Sea of Tranquillity in which Apollo 11 landed was pictured by Apollo 10. Eagle landed in a spot close to the shadows at the top left of the image.

One Small Step

Their nervous energy sapped by the landing, TV viewers had time to recover before the historic moment—the first time that an Earthling was to set foot on another world. Late on July 20th U.S.-time, *Eagle* was depressurized and the hatch opened. Neil Armstrong slowly backed out of the hatch onto the ladder, and down nine rungs to history.

Below: Buzz Aldrin features in one of the many superb images taken by Neil Armstrong. One of the footpads of the *Eagle* lunar module can be seen.

On the way down, he pulled out a lanyard which exposed a TV camera, which beamed live black and white images to Earth—although at first they were upside down. On his way down, Armstrong jumped on the footpad and tested his ability to jump back up onto a rung.

He paused and said, "I'm going to step off the LM now." Placing his right boot into the soil, he said, "that's one small step for man, one giant leap for mankind." Ironically, it is one of the most misquoted quotations in history. He never did say "a man."

Aldrin came out, and on stepping onto the moon, said "magnificent desolation." They got to work, putting up the U.S. flag, preparing to lay out some instruments on the surface, and taking some vital rock and dust samples. Armstrong proved to be the best lunar photographer, taking images of Aldrin, one of which became the symbol of the

flight and whole Apollo enterprise. There are no formal stills of Armstrong on the moon because Aldrin did not take any, except one accidental one when he was taking a 300° panorama in which Armstrong could be seen in shadow and with his back to the camera.

President Nixon got in on the act with a gushing speech: "This has got to be the most famous call ever made," he said. Kennedy's goal had been met. The race to the moon had been won.

The 2 hour 21 minute moonwalk ended and the ascent stage took off, rendezvoused, and docked with *Columbia* for the journey home. After splashdown the astronauts were placed in a quarantine container, in case of possible lunar bugs. Then the three astronauts faced a frenzied welcome and months of appearances at palaces, press conferences, and official dinners.

Storms ahead

Next to go to the moon were the Apollo 12 crew, who were launched on November 14, into thunderous rainclouds to please President Nixon with an on-time lift-off during his visit to the Kennedy Space Center. It nearly proved a disastrous decision because at T+36 seconds, with the Saturn enveloped in clouds and out of sight, it was struck by lightning, knocking out the command module systems. The test pilot commander, Pete Conrad read out the failures, "we just lost platform gang... we had everything in the world drop out... fuel cells, lights, and AC bus overload 1 and 2... main bus A and B...".

Miraculously, a flick of a switch on command

from mission control solved the problem, and Conrad and his companions Alan Bean and Dick Gordon headed, fittingly, toward the Ocean of Storms for a pinpoint landing just 180 meters from the Surveyor 3 soft-lander which had arrived in 1967. On November 20, Conrad and Bean visited Surveyor 3 and removed about 22 lbs (10 kg) of parts from the spacecraft, including the TV camera, for later examination back on Earth. The Surveyor 3 camera is now on display in the Smithsonian National Air and Space Museum in Washington, DC.

A humorous moonwalk by Conrad and Bean was broadcast to Earth, until the TV camera failed. TV audiences were reduced to listening to the two moonwalks of the mission as if on a radio. This was a disappointment. Already, with the goal of the moon won, public apathy was setting in, and the political will to continue with Apollo was waning fast.

Left: A cine camera image showing Neil Armstrong's first steps. Notice the stark shadows on the surface. Aldrin did not take any formal images of Armstrong.

Below: Apollo 12's Pete Conrad is pictured by Al Bean with the Surveyor 3 spacecraft, which soft-landed in April 1967 on one of the many reconnaissance and technology demonstration missions before Apollo 11.

"Houston, We've Had a Problem Here"

Launched routinely on April 11, 1970, Apollo 13 received far less public attention than its predecessors. Fifty-six hours later that was all to change, heralded by Jack Swigert's famous words. The three crewmen were now in grave danger of becoming the first Americans to perish in the freezing vacuum of space.

The crisis seemed to come out of nowhere. On April 13, on the mission's designated flightpath toward the Fra Mauro highlands, James Lovell, Jack Swigert, and Fred Haise had just finished a TV broadcast from the command module *Odyssey*, that unbeknown to them, no U.S. network had used. Inside the service module was a time-bomb, a faulty oxygen tank in the fuel cell system which provided electricity. Heater switches had been welded shut during a ground test and not checked, and when the crew was asked to "stir up the cryo tanks," the inevitable explosion happened.

There was a loud bang and shudder. Swigert said quietly, "Houston, we've had a problem here." Houston didn't catch the urgency. Then Lovell came on the line, "Houston, we've had a problem. We've had a main B bus undervolt." With the fuel cell system damaged and electrical power dwindling, the command module would soon become useless.

The moon landing was off, and a drama unfolded that captured the attention of the world, as engineers on the ground worked out ways to keep the crew alive and get them back safely. It is ironic that because of this crisis, it is Apollo 13 that most people now remember, not Apollo 11.

The crew was fortunate that the accident occurred when it did, because the lunar module Aquarius was still attached and could be used for engine firings to get the crew on course for Earth and to help provide life support.

After receiving advice from mission control and the private contractors who had built the spacecraft, the Apollo 13 crew made three burns of Aquarius's engine, taking the crew around the moon and back to Earth while retaining life support. One of the life-saving steps taken was to create a carbon dioxide filter from spacesuits intended for moonwalking, and other non-essential equipment. Despite these measures, the conditions aboard became pitiful and the crew was very fortunate to have survived.

Right: The Apollo 13 command module *Odyssey* is safe at last on board the aircraft carrier recovery ship.

The world held its breath for four days during the Apollo 13 saga, and people were overjoyed by its safe return. President Nixon eloquently summed up the team effort, echoing Winston Churchill: "never have so few owed so much to so many."

Curtailing the dream
By 1970 President Nixon knew that support for Apollo was on the wane and draining America's resources. He canceled Apollos 18–20. Just four Apollo missions were left, and each was under

threat of further cuts.

Apollo 14 was aimed at the Fra Mauro highlands that Apollo 13 had been earmarked to explore. Its commander was to be none other than America's first astronaut Alan Shepard, who had been grounded cruelly by an inner ear problem for many years. Apollo 14 was launched on January 31, 1971, and almost immediately experienced difficulties, with persistent failures to dock with the lunar module encased on the S4B third stage after the translunar insertion burn. The docking was finally successful and Shepard eventually landed his *Antares* lunar module, but only after a radar problem nearly aborted the approach to Fra Mauro.

As he stepped out onto the surface of the moon, shielding his eyes from a very low sun, Shepard said, "it's been a long way but we're here," alluding in part to his long battle to be restored to flight status. He and his pilot Edgar Mitchell laid out a suite of scientific instruments but had difficulty reaching their target of Cone Crater. The surface was more undulating and covered with more boulders than expected, which disorientated the astronauts.

With two moonwalks completed, the crew headed for the command module Kitty Hawk, piloted by Stuart Roosa. On their arrival back home, they were the last to have to endure the quarantine container.

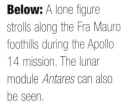

Left: President Richard Nixon (right) greets the Apollo 13 crew after their landing. Left to right are Jack Swigert, Fred Haise, and James Lovell.

Below: A lone figure strolls along the Fra Mauro foothills during the Apollo 14 mission. The lunar module *Antares* can also be seen.

"Exploration at its Greatest"

The astronauts on the first three expeditions to land on the moon were restricted to a safe walking distance from the lunar module in case of trouble, but the Apollo 15 moonwalkers had the luxury of being able to extend lunar exploration from yards to miles, with a Lunar Roving Vehicle (LRV). As if that was not enough, Apollo 15 was targeted at one of the most spectacular areas of the moon, on the edge of the Mare Imbrium, next to the Apennine Mountains and close to the Hadley Rille.

It all made for what has been described by NASA as the "most complex and carefully planned scientific expedition in the history of exploration." The mission was an all-U.S. Air Force affair, led by Colonel Dave Scott. His lunar module (*Falcon*) pilot was James Irwin and the command module pilot, Al Worden.

Launched on the seventh manned Saturn 5 booster on July 26, 1971, the flight to the moon was largely uneventful. On its final approach, the *Falcon* made a steep 26° descent over the Apennine Mountains and landed safely near the edge of Hadley Rille, and as expected the view was a fantastic sight for Scott and Irwin.

After exiting *Falcon*, the ingeniously-designed LRV was unfurled from the side of the lunar module and unfolded to form what looked like a dune buggy. In fact it was a highly sophisticated vehicle designed especially to operate in a vacuum, in wide extremes of temperatures, and over difficult terrain. It was 10 ft (3 m) long and 7 ft (2 m) wide,

Below: One of the classics of Apollo: David Scott salutes the flag at Hadley Base, with the smooth mountains of the moon in the background.

and had four wire-mesh wheels. Powered by two silver zinc batteries, the rover was equipped with a high gain antenna providing direct communications with the Earth.

The ultimate drive

A TV camera mounted on the rover provided Earth viewers with spectacular rides around Hadley Base on three EVAs, lasting over 18 hours. "This is exploration at its greatest," chortled Scott during one of the moonwalks. The LRV had traveled just over 17 miles (27 km) when it was parked for a final time close to the lunar module, with the TV camera ready to cover the lift off of the ascent stage "live" for TV views.

The stage—with a cargo of 173 lb of moon rocks and Scott and Irwin—took off in a shower of multicolored sparks, ending a breathtaking exploration. On the way home, the command module pilot, Al Worden, made a unique EVA 199,000 miles from Earth, to retrieve experiment packages from the side of the service module.

While Scott and Irwin were quite business-like in their exploration, the next pair of moonwalkers on Apollo 16 were like a comedy duo. Apollo 16 was aimed at a highland area called Descartes and was launched on April 16, 1972, the beginning of the end for Apollo because missions 18-20 had been canceled due to budget cuts and public apathy. Only Apollo 17 was left.

The usually taciturn veteran astronaut John Young and the rookie Charlie Duke (previously Mission Control Capcom), with an infectious enthusiasm, made their three moonwalks fun to

listen to. Their banter included jokes, impersonations, and moon jumps, during one of which Duke nearly killed himself by going off-balance in mid-air and landing back on the moon on his life supporting backpack.

Without realizing it and because he could not see his feet in the bulky suit, Young tripped over a wire of a heat flow experiment and ruined it. The experiment was one of a suite of experiments laid out on the surface, similar to those left by previous Apollos.

The Apollo 16 duo rode the LRV during three moonwalks lasting over 20 hours, and during one drive Young drove the LRV at 8 mph (13 km/h), churning up the moondust.

On the way home, the CMP Ken Mattingly made a trans-Earth spacewalk. Apollo 16 brought another 213 lb (96 kg) of moonrocks to Earth following an 11-day journey.

Above: Jim Irwin unloads the lunar rover during one of the three Apollo 15 moonwalks.

Left: Apollo 16's Saturn 5 rises from the lush landscape of the Kennedy Space Center.

Below: John Young jumps in one-sixth of the Earth's gravity at Descartes during one of the Apollo 16 moonwalks. Charlie Duke took the picture.

Technological Triumph

On December 7, 1972, a Saturn 5 launch with a difference took place at the Kennedy Space Center as Apollo 17 blasted off in the dead of night. The start of the Apollo 17 mission was a spectacular affair, turning night into day. The objective of the flight: to land on the Taurus Littrow highlands.

Below: The stark beauty of the desolate lunar landscape can be seen in this image of Jack Schmitt working on the rover at Taurus Littrow.

The mission commander was Gene Cernan and his lunar module pilot was Jack Schmitt, a professional geologist, with the command module being piloted by Ron Evans. Cernan and Schmitt flew the lunar module *Challenger* to a perfect landing.

During three moonwalks, using the third and last lunar rover, the crew covered a lot of ground and discovered that some of it was orange. This exciting discovery was interpreted by Schmitt as evidence of recent volcanism, and by others as pointing to evidence of water under the surface. It had made sense to send a geologist to the moon, rather than a test pilot. Harrison "Jack" Schmitt was originally earmarked to fly Apollo 18, but due to the budget cuts there was no longer going to be an Apollo 18, so Schmitt replaced test pilot Joe Engle. Engle lost the moon.

Since Apollo 15, the test pilot crews had all been well-trained in geology, and on Apollo 17 the lunar module crew effectively comprised a pilot-geologist and a geologist-pilot! Schmitt was a worthy addition to Apollo 17, but the time on the moon was limited so it was impossible to do all the research hoped for.

For all mankind

Before he left the moon, hoping he would not be the last lunar astronaut, Cernan said; "As I take these last steps from the surface, back home for some time to come, but we believe, not too long into the future, I believe history will record that America's challenge of today has forged man's destiny of tomorrow. And as we leave the moon and Taurus Littrow, we leave as we came, and God willing, as we shall return, with peace and hope for all mankind."

In many of his stirring speeches before Apollo 17, commander Gene Cernan said his mission was just "the end of the beginning," but many thought the opposite. It was the beginning of the end. Realistically, whatever Cernan might have hoped, the dreams of further flights to the moon had died under the budget ax, and there were no serious plans to return to the moon. Following six successful landing missions, there seemed no need to return for decades to come.

Wernher von Braun said: "Apollo cannot be revived. Most people have left and we have passed the point of no return... the awareness of the dream has gone sour, and a great journey has petered into something of a dead end."

The Apollo program rushed to the moon as quickly and simply as possible in a political tour de force that was described by one newspaper as "a technological orgasm." Apollo had achieved everything that John F. Kennedy—and the American people—had hoped for. The $25 billion cost of Apollo has often been criticized, yet this sum is small compared with what Americans spend on their pets alone.

Why was it the end? "Maybe the answer lies in the psychology," wrote British space correspondent Angus McPherson at the time. "The gasp of wonderment and awe of the watching human race had frozen into an irritable yawn in four years, since Apollo 8's epic mission—which was for some the highlight of space exploration, rather than Apollo 11."

Left: Gene Cernan is seen riding the lunar rover during one of the three spacewalks by the last lunar explorers.

Below: The end of the beginning... or just the end? Apollo 17 returns with a splash.

Science and Spin-off

Many direct advances in science were made as a consequence of Apollo, but more subtle advances were also made which were of more use to society in general. These include the development of everyday items such as computers and power tools—pioneered by scientists at NASA and by the private contractors who built the Apollo program.

The six Apollo lunar landing missions returned 850 lbs (385 kg) of moon rock to the Earth. Total moonwalk time amounted to three days, eight hours, and 22 minutes, in 14 sorties, and lunar surface stay time was 12 days, 11 hours, and 40 minutes. Each landing mission laid out a collection of scientific instruments on the surface, many of which operated after the astronauts had left the moon. These were Apollo Lunar Surface Experiments Packages (ALSEP), which included seismograph, dust detection, a magnetometer, spectrometer, ion detector, cathode gauge, charged particle detector, a heat flow experiment, gravimeter, atmospheric composition, and ejecta and meteorite detectors.

Apollos 15–17, in particular, carried out extensive lunar observations from orbit using a suite of instruments operated by the command

Above: One of the Apollo 11 moonrocks in a chamber in the Lunar Receiving Laboratory at Houston.

module pilot. These included a camera derived from spy-satellite technology capable of taking images of the surface to a detail of 3 ft (1 m) across. Including Apollos 8 and 10, the total amount of time in lunar orbit amounted to 27 days, 17 hours, and 46 minutes—or 363 orbits in total; 24 people traveled around the moon or orbited it, and 12 walked on its surface.

What was learned from the Apollo program? First, the moon is not a primordial object; it is an evolved terrestrial planet with internal zoning similar to that of Earth. We now know that the moon is made of rocky material that has been variously melted, erupted through volcanoes, and crushed by meteorite impacts. It is an ancient body which formed at the same time as the Earth, and the interpretation of other planets is based largely on lessons learned from the moon.

Before Apollo, the origin of lunar impact craters was not fully understood, and the origin of similar craters on Earth was highly debated. The youngest moon rocks are about the same age as the oldest rocks found on Earth. The earliest processes and events to have occurred on both planetary bodies can now only be found on the moon. Scientists believe that moon rock ages range from about 3.2 billion years in the dark, low basins called maria, to nearly 4.6 billion years in the light, rugged highlands.

An explosive history

The moon and Earth are very much related, but the moon is lifeless; it contains no living organisms, fossils, or native organic compounds. All moon rocks originated through high-temperature processes with little or no involvement with water. They are roughly divisible into three types: basalts, anorthosites, and breccias.

Early in its history, the moon was melted to great depths to form a magma ocean. The lunar highlands contain the remnants of early, low-density rocks that floated to the surface of the magma ocean. The lunar magma ocean was followed by a series of huge asteroid impacts that created basins that were later filled by lava flows.

The large dark basins, such as Mare Imbrium, are gigantic impact craters, formed early in lunar history, that were later filled by lava flows about 3.2–3.9 billion years ago, say geologists. Lunar volcanism occurred mostly in the form of lava floods that spread horizontally; volcanic fire fountains produced deposits of orange and emerald-green glass beads.

The moon is slightly asymmetrical, possibly as a consequence of its evolution under Earth's gravitational influence. Mass is not distributed

uniformly inside the moon. Large mass concentrations, called mascons, lie beneath the surface of many large lunar basins and are probably caused by thick accumulations of dense lava. The surface of the moon is covered by a rubble pile of rock fragments and dust, called the lunar regolith, that contains a unique radiation history of the sun that is of importance to understanding climate changes on Earth.

Apollo also kick-started a technology revolution that continues today. Called space spin-off at the time, the technological innovations and developments, particularly in computers, result from space age developments. From a huge computerized instrument unit on a Saturn 5 with the power of one of today's simplest pocket calculators, that technology has advanced to the almost unbelievable and rapid communications of the Internet. Similar spaceflight technology spin-offs range from hi-tech surgical instruments and power tools to fire proofing materials. However, the much-quoted non-stick frying pan was not one of the Apollo spin-offs.

Left: Jack Schmitt—the only professional geologist to walk on the moon—uses a rake tool to sieve moondust at Taurus Littrow.

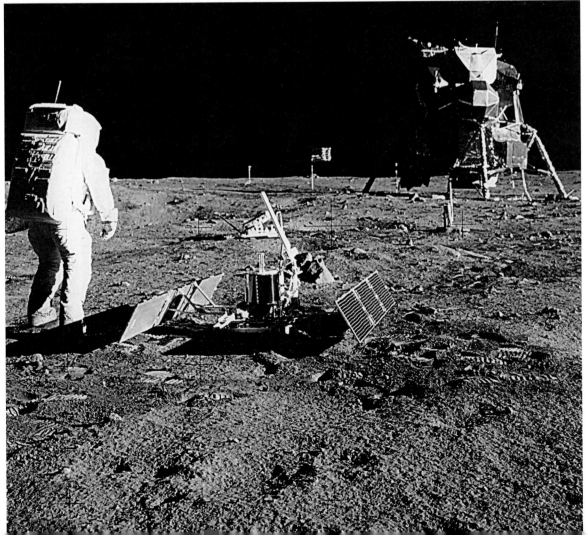

Left: Buzz Aldrin alongside the Passive Seismic Experiment at Tranquillity Base.

What Happened to the Russians?

The Soviet Union lost the race to the moon in 1968. Unlikely to land on the moon before America, it had hoped to fly two cosmonauts around the moon to steal some thunder. It was the fear of this that led NASA to launch Apollo 8 to orbit the moon in December 1968. This, however, was the last time that the Soviet Union would be in a position to compete with the United States' lunar program.

This Soviet lunar-loop program was called Zond, or L-1. Zond was based on the new Soyuz manned spacecraft, built primarily as a space taxi for a planned space station program. The Soyuz, without an orbital module and capable of flying only two crew, was used for the Zond program with a relatively new Proton booster.

After some Earth orbit tests, the first lunar Zond was launched as Zond 4 (there had been previous planetary missions) on March 2, 1968. The unmanned craft flew around the moon but was deliberately destroyed on its return journey as it was heading for a landing outside Soviet control.

Zond 5 flew around the moon and landed in the Indian Ocean after a control failure placed the craft on such a steep trajectory into the Earth's atmosphere that a crew would have been killed. In November 1968, Zond 6 flew around the moon but depressurized during the return to Earth. The parachute also failed and the craft was destroyed. Both malfunctions would have killed a crew.

Zond 7 was the only completely successful mission. The craft looped the moon and returned safely to the Soviet Union in August 1969. Zond 8 also made a lunar loop in October 1969 but this time a control system failure resulted in another landing in the Indian Ocean.

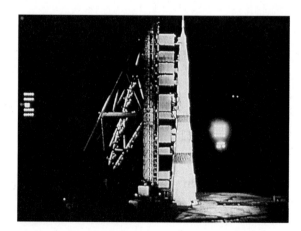

Disaster after disaster
The T-3 manned lunar landing program, meanwhile, ended in a complete disaster, with four consecutive failures of the N-1 boosters. The first N-1 giant booster, carrying a simulated moon lander, rose from its Baikonur launch pad on February 26, 1969. At T+66 seconds, an oxidizer

pipe to one of the 30 first-stage engines ruptured and leaked liquid oxygen, which caught fire. The engines were shut down by the flight computer and the launch escape system fired, carrying the lander to safety, before an enormous explosion.

One July 3, 1969, a metallic object fell into the oxidizer pump of engine No. 8 of the second N-1. The engine exploded, disabling other engines and control cables, and the vehicle fell back onto the pad and exploded, destroying not only its own launch pad but an adjoining N-1 launch complex.

N-1 No. 3 lifted off from the repaired adjoining launch pad, on June 27, 1971, carrying a mock-up of the entire moon lander system, but almost immediately experienced roll problems, and by T+39 seconds the roll had exceeded the limits of the launcher's control system. At T+48 seconds, the N-1's second stage started to break apart, and at T+51 seconds the automatic flight control system shut down all the engines. Yet another N-1 plunged into the Baikonur steppe land.

On November 23, 1972, the final N-1 was launched with a similar payload to N-1 No.3, and at T+90 seconds the planned shutdown of central engines of the N-1 first stage caused an overload of

pressure which ruptured propellant lines, and the vehicle caught fire. About 20 seconds later, the first stage exploded.

In the end all the Soviets got was about 4 oz (113 g) of moon. The first return to Earth of moondust was achieved by the Soviet Union, using an unmanned spacecraft, Luna 16, which was also the Soviets' true soft-lander. Luna 16 landed on the moon in September 1970 and incorporated a lunar collector, which scooped up some soil and deposited it inside the capsule on top of an ascent stage. The 1.63 ft (0.5 m) diameter capsule, weighing 23 lb (10.5 kg), eventually parachuted to Earth carrying 2.64 oz (75 g) of lunar material from the Sea of Fertility. By this time, of course, Apollo missions had returned with pounds of rock.

Two further Soviet sample returns were performed in 1972 and 1976, but by this time such missions were of little scientific value.

The Soviet N-1 booster, seen **far left** being rolled out to the pad, **left** erected on the pad at Baikonur and launched, had a wretched flight history, including the loss of Flight 3 in June 1971, **inset** and **right**.

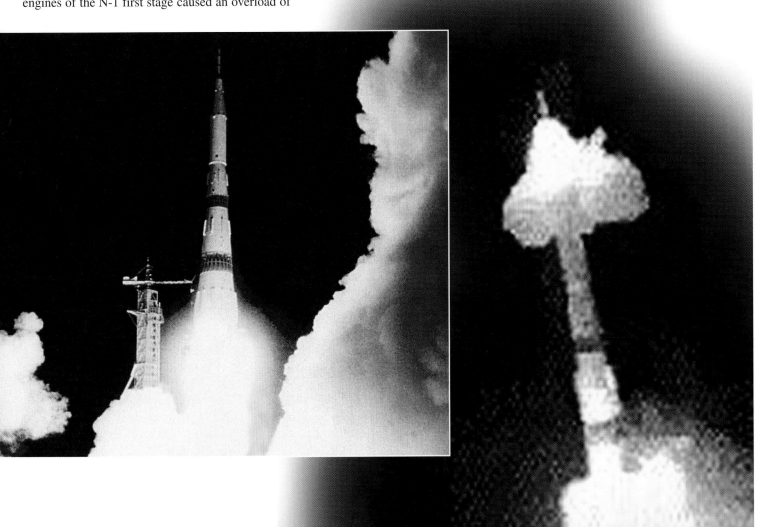

CHAPTER SIX

SPACE SHUTTLE

The Birth of the Space Shuttle

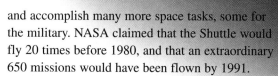

Probably the most famous space vehicle today, the Space Shuttle has flown over 100 missions since 1981 and has made space travel look almost routine, though by no means the frequent occurrence that was hoped for by its visionaries.

Many artists' early concepts of space travel of the future were entirely logical, featuring winged, airplane-like rocketships. If there had not been a Cold War, perhaps such vehicles would have evolved sooner.

In early 1970, a buoyant NASA's next goal was to build a space station and use a "taxi" to ferry cargo and crew to and from it, making space travel much more routine. The White House and Congress, however, canceled some later Apollo missions along with the space station project.

NASA was left with the space taxi—with nowhere to go. So, the taxi was renamed the Space Shuttle by NASA and advertised as a fully reusable, versatile system that would carry commercial satellites into orbit, charging a fare, and would act as a mini-space station and laboratory. It would also carry out repairs in space,

Below: The reusable two-part system (right) compared with a spaceplane flying on a disposable Saturn 5 first stage.

and accomplish many more space tasks, some for the military. NASA claimed that the Shuttle would fly 20 times before 1980, and that an extraordinary 650 missions would have been flown by 1991.

But the budget for the Space Shuttle was the equivalent of about one fifth the cost of Apollo—to build a system that would be even more challenging than Apollo had been. The result was inevitable. The design chosen was, as a matter of financial necessity, not fully reusable and basically an engineering compromise.

Transporter and science lab

The Space Shuttle comprises three main elements. The first is the orbiter, which is equipped with three main engines called the SSMEs, which are fueled by liquid oxygen and liquid hydrogen fed from a large brown external tank (ET) attached to the belly of the orbiter. Attached to either side of the ET are two solid rocket boosters (SRBs) which supplement the SSMEs during the first two minutes of flight, and which are recovered from the sea after each flight. Most parts of the SRBs are used again on future Shuttle flights.

The Shuttle's payloads are carried primarily in the 60 ft (18.3 m) long, 15 ft (4.6 m) wide payload bay. Additional cargo and experiments, as well as crew equipment and consumables, are located in the mid-deck which is positioned under the flight deck. The mid-deck also acts as the wardroom, kitchen, gym, toilet, bedroom, and as an airlock, used for EVAs and as an entrance to the attached Spacelab and Spacehab modules.

Spacelab is a laboratory which is fitted inside the payload bay and which has been used many times for various science missions. Spacehab provides additional working, storage, and instrument space for some Shuttle missions.

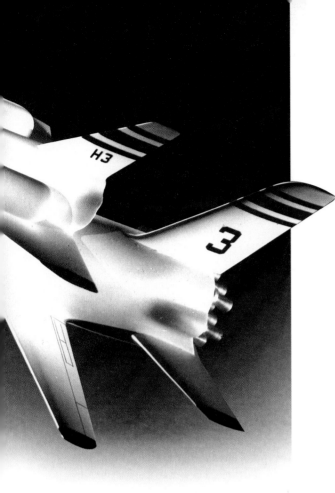

Like Spacelab, it extends into the payload bay.

The original maximum payload carrying capability for a 28.5° inclination low-Earth orbit by the Shuttle was advertised as 65,000 lb (29,483 kg), but the heaviest payload so far carried weighed 52,300 lb (23,723 kg)—on the ill-fated *Challenger* which was lost during launch in 1986. The heaviest payload since then was the 49,789 lb (22,584 kg) on STS-93 *Columbia* carrying the Chandra X-ray telescope in 1999.

The reduction in payload carrying capability has been the result of an increase in the weight of the total Shuttle system, mainly as a result of redesigns to strengthen the whole system during launch, made after the *Challenger* accident. The *Challenger* disaster exposed not just ill-designed SRBs but launch dynamic loads far in excess of those predicted when the Shuttle was designed—a fact that has been extremely well-hidden by NASA.

A record 17-day mission was made possible by the use of a relatively new unit called the Extended Duration Orbiter (EDO) unit, which adds more fuel cells to the orbiter, for additional electrical power.

Left: The ideal design of the U.S. Spaceplane would have been a combination of a reusable booster and spaceplane, both of which would fly back to base like an aircraft.

Below: The Space Shuttle orbiter, seen here in the background of a NASA conceptual painting, was to have made 650 flights by 1991, mainly servicing the planned Space Station.

The System

Incorporating features of a spacecraft and airplane, the Space Shuttle is still the world's only re-usable space vehicle. Despite the design now being over 20 years old, many of its components—including on-board computers and cockpit controls—are continuously upgraded using the very latest technology.

The Space Shuttle orbiter is 122.2 ft (37.24 m) long with a wingspan of 78.06 ft (23.79 m). The total length of the whole stack from the tip of the external tank (ET) to the tail of the solid rocket boosters (SRB) is 184.2 ft (56.14 m).

The Space Shuttle on the launch pad and fully loaded for lift-off weighs about 4.5 million pounds (2,041,000 kg), of which the orbiter accounts for about 250,000 lb (113,400 kg).

The SRBs are the largest solid-propellant motors ever flown and the first designed to be re-used. They each weigh about 1,259,000 lb (57,107 kg) at launch, 85% of that being the propellant. The SRBs develop a thrust of about 3,300,000 lb (1,496,870 kg) each, and only ignite after the liquid oxygen-liquid hydrogen Space Shuttle Main Engines (SSMEs) have built up to full thrust. They provide 71.4% of the thrust of the vehicle at lift-off. The reusable SRBs burn for 120 seconds, burn out, and are jettisoned, splashing down in the Atlantic Ocean under three parachutes, about 141 miles (225 km) downrange, approximately 281 seconds after launch.

The ET can now weigh as little as 1,648,433 lb (747,475 kg) with the use of an aluminum-lithium alloy and lighter foam for the tank's insulaton. The ET remains attached to the Space Shuttle until the cut-off of the SSMEs, which occurs at about 59 miles (95 km) altitude, and is jettisoned to burn up in the Earth's atmosphere.

The reusable SSMEs can be gimbaled for pitch, yaw, and roll control, and are throttleable within a range from 109% rated thrust, a reduced 60% during Max Q at about T+60 seconds, and later the engines throttle up again. The SSMEs have been improved significantly during the lifetime of the Shuttle program to reduce wear and tear on components.

After SSME shutdown, two Orbital Maneuvering System (OMS) engines, mounted on pods on either side of the tail, are fired, sometimes several times, to accelerate the spacecraft to orbital velocity. The OMS are also the Shuttle orbiter's retro rockets, and reduce the orbital speed by 300 ft/s (91 m/s) during a two-minute burn.

Yaw, pitch, and roll maneuvers and small velocity changes are performed by the orbiter's Reaction Control System (RCS), which comprises

38 primary thrusters, 14 on the nose and 12 on each OMS pod, and six small thrust verniers, two in the nose and four at the rear.

Flight management

It is impossible to fly the orbiter without computer assistance. Four computers operate in parallel during critical flight operations such as launch, ascent, de-orbit, re-entry, and landing, "voting" on every input and response 440 times a second as a safeguard against an individual computer failure. More up-to-date units are being used with the introduction of glass cockpit displays to eventually replace the original switches and dials on all orbiters.

Most flights of the Space Shuttle carry the 50 ft (15.24 m) long Remote Manipulator System (RMS), which is a sophisticated robot arm controlled from the flight deck by specialist RMS operator astronauts. At the end of a mission, after the OMS retrofire, the Shuttle starts re-entry at about 400,000 ft (121,920 m), approximately 5,000 miles (8,000 km) from the landing site at a speed of about Mach 25. It can make maneuvers during re-entry to enable it to fly 750 miles (1,200 km) either left or right to align itself for an emergency landing.

The heat builds up due to the friction against the atmosphere, and unlike the ablative heatshields on previous manned spacecraft, the Shuttle is equipped with unique thermal tiles and blankets. The are six types of this material because different parts of the orbiter experience varied heat levels, the wing leading edges and underside bearing the brunt of the re-entry heating, while the upper part of the fuselage is less exposed to heat.

The unpowered gliding approach to the landing site, usually at the Kennedy Space Center, is made at seven times steeper an angle and 20 times faster than an airliner flies, and the speed at touchdown is over 200 mph (320 km/h). A drag chute—introduced for flight STS-50—is deployed to aid braking.

Far left: The reusable segmented Solid Rocket Booster was the cheaper replacement for a planned safer and more controllable liquid propellant booster.

Left: The Shuttle orbiter *Discovery*, with its payload bay doors open, to reveal the empty area previously occupied by a satellite that has been deployed. This image was taken by a free-flying experimental platform also deployed from the payload bay.

Below: The External Tank pictured after the Shuttle has entered a preliminary low orbit. The ET plunges into the atmosphere and is destroyed.

Shuttle Test Flights

Since 1946, American X-planes had been demonstrating unpowered returns to Earth following rocket-powered ascents to the edge of space. Knowledge gained from these flights would be crucial to the development of the Space Shuttle.

Below: Pilot Bill Dana pictured after a HL-10 lifting body flight, looking up at the B52 mother plane which had deployed him and his craft at high altitude.

The final X-15 mission was flown in 1968. From 1963–75, rocket-plane research was complemented by wingless lifting-bodies, aircraft that could fly using aerodynamic lift and high lift-to-drag ratios to make steep descents with quite acute maneuvers. The M-2 series, nicknamed the flying bathtub, and later the HL-10 and X-24, performed a series of simulated returns of an unpowered shuttle vehicle from space.

Space Shuttle astronauts were the first to testify to the enormous debt that the Shuttle owed to these rather unsung programs, and their pilots who included legends such as Milton Thompson, Bill Dana, and Bruce Petersen, whose horrific M2/F2 crash, which he survived, was immortalized by the TV series *The Six Million Dollar Man*.

Once the Space Shuttle had been fully designed and the first model, *Enterprise* built, five approach and landing tests (ALT) were made between August and October 1977 by Shuttle astronauts, to fully simulate the descent and landing phase of the mission.

Five ALT free-flight missions were flown by two teams led by Fred Haise, the Apollo 13 veteran, and Joe Engle, the former X-15 rocket plane pilot. The *Enterprise* was flown on top of a modified Boeing 747 and released to allow the pilots to practice the 220 mph (354 km/h) landings.

All was ready for the first space mission, originally scheduled for 1978, by the first space-worthy Shuttle orbiter, *Columbia*. However, this was delayed until April 12, 1981 by several technical problems that were beginning to give the Shuttle a bad name.

Persistent problems

Shuttle pioneers John Young—the first over-50 in space—and Bob Crippen flew a near-flawless two-day mission, making an epic landing at Edwards Air Force Base, California. During the mission, there had in fact been enough concern about the loss of some heatshield tiles for a spy satellite to be used to take images of the Shuttle's

underside, to ensure that none of the most critical tiles had been lost.

NASA was also concerned about damage to the launch pad, including a huge crack on the mobile launch platform. Perhaps the dynamics loads at lift-off were greater than anticipated. NASA did not allude to this possibility, but judging from what happened on later missions, clearly something had been underestimated. An improved sound-suppression system was introduced for the next *Columbia* test flight to reduce pressure pulse.

The launch was made on November 12, 1981 with Joe Engle and Dick Truly aboard, but they had to be brought home after just two days into a five-day mission after a fuel cell failure. Still, the crew had an opportunity to give the Remote Manipulator System (RMS) a work-out.

STS-3, launched on March 22, 1982—with a brown external tank, rather than white to save the cost and weight of the paint—lasted eight days and demonstrated the use of the RMS with simulated satellite deployment and retrieval. The mission was packed with dozens of experiments but there were more niggling failures—of the toilet, an APU, a TV camera, and the radio. In addition, a payload bay door froze and *Columbia* lost some more tiles. The landing was delayed due to high winds at Edwards, so *Columbia* was aimed to a stand-by site at White Sands, New Mexico.

The fourth and final test flight was made on June 27, 1982, and carried some Department of Defense payloads. The two crew were greeted by an ebullient President Ronald Reagan who declared the Space Shuttle operational.

The Pressure Mounts

In 1982, NASA was under pressure to justify the expense of the Space Shuttle with regular, routine missions. The agency even introduced a new mission for the Shuttle—it was also to be a commercial launch vehicle. The U.S.'s expendable launch fleet would be phased out, and the Shuttle would provide America's sole launch capability.

The pressure on the system and launch schedule was going to be intense. The first operational mission of the Space Shuttle, STS-5/*Columbia* was launched on November 12, 1984, and carried two commercial communications satellites with their own upper stages to enable them to eventually get into geostationary orbit.

Carrying four crew for the first time, including two new-class astronauts called mission specialists, with no ejection seats for emergency escape, *Columbia* deployed the satellites, but the first Shuttle spacewalk had to be canceled.

In 1983, STS-6 deployed the first of NASA's fleet of Tracking and Data Relay Satellites, but this was stranded in the wrong orbit due to an upper

Below: A commercial communications satellite is deployed from the Space Shuttle during one of many commercial missions, which were eventually discontinued.

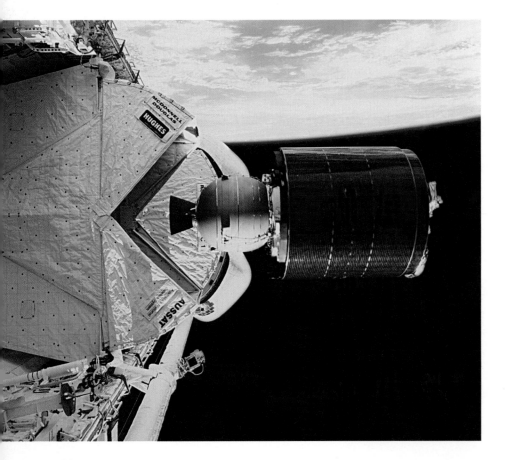

stage failure. Mission seven carried five crew including the first American woman in space, Sally Ride.

The ninth mission carried the first Spacelab laboratory designed for intensive science research. A new breed of non-NASA payload specialists was introduced, including a German national, and six people were put in space.

Salvaging space debris

In 1984, on another trucking mission, STS-41B, two deployed satellites were left stranded in orbit but two astronauts successfully tested a new manned maneuvering unit (MMU) to allow them to fly independently in space. The first was Bruce McCandless. This tenth mission was also the first to land back at its launch base, the Kennedy Space Center. The 11th mission performed the first space repair of a retrieved satellite when two astronauts mended the Solar Max satellite and redeployed it into space. The Shuttle was certainly proving to be a vastly capable and versatile vehicle.

The twelfth mission carried the first industry astronaut, Charlie Walker from McDonnell Douglas, who was flying to operate a new device to possibly manufacture ultra-pure pharmaceuticals in the weightlessness of space. Mission 13 carried seven people, and space was now being touted as the new commercial frontier, with the Shuttle as the space truck.

Another 1984 mission, STS-51A, performed the greatest feat to date by capturing the two satellites lost by STS-41B and bringing them back to Earth. Astronauts Joe Allen and Dale Gardner flew MMUs and grabbed the satellites using special devices, and brought them back to the payload bay in a remarkable and spectacular display of space salvage. Both satellites were later re-launched on expendable launchers.

A classified military mission was flown on the first 1985 sortie, while mission 16 launch in April, in questionable weather, carried a Congressional observer, Senator Jake Garn. It also lost another satellite which was to be retrieved, repaired, and redeployed later in the year by another Shuttle mission.

A Spacelab mission, STS-41F in July 1985 flew the first ascent abort mode in the program, when one engine shut down prematurely. Orbit was achieved, however. Later in 1985, the first—and so far only—eight person flight was launched on a Spacelab mission.

The schedule for 1986 had 15 missions designated, with over 15 commercial satellites and an extraordinary dual mission with two launches in six days in May to deploy two planetary spacecraft, *Galileo* and *Ulysses*, using a fully liquid-fueled upper stage called Centaur G. The risks involved were illustrated by the plan to vent the cryogenic propellants over the wings of the Shuttle in the event of an abort.

The first launch from Vandenberg Air Force Base, California was also scheduled on a military mission. The Department of Defense had big plans for the Shuttle. Incredibly, the U.S. Air Force Under-secretary was planning to hitch a ride on this maiden flight!

More commercial satellite customers were negotiating flights of company specialists aboard. NASA was keen to accommodate any passenger. Two other 1986 flights were to carry a teacher and a journalist—most likely Walter Cronkite—respectively. The pace was manic, and the NASA hype extraordinary. It was obvious to seasoned space observers that things were getting out of hand. The Space Shuttle just could not keep up with this pace. Something had to crack.

Above: An astronaut makes an independent EVA flying a capture mechanism, to snare a communications satellite stranded after a Shuttle deployment and bring it back to Earth.

Left: Astronauts at work aboard the German-funded Space Shuttle Spacelab mission in October-November 1985, in which a record eight-person crew was launched for the only time in space history.

The "Major Malfunction"

There was a feeling that NASA was flying close to the wire even before STS-51L was launched. After several delays, the crew climbed aboard *Challenger* on the very cold morning of January 28, 1986. Rookies, pilot Mike Smith, company specialist Greg Jarvis, and President Reagan's Teacher-in-Space finalist, Christa McAulliffe, were joined by veteran crewmembers, commander Dick Scobee and mission specialists Ronald McNair, Judy Resnik, and Ellison Onizuka. American eyes were on Christa McAuliffe, first teacher in space.

At T+73 seconds after the launch into the cold clear skies of the Kennedy Space Center, *Challenger* disappeared in an explosion. Launch commentator Steve Nesbitt spoke the words, "obviously a major malfunction," as people looked on in disbelief and shock.

As America grieved and divers searched for the bodies of the seven in the shattered cockpit of *Challenger* under the Atlantic Ocean, the Rogers Commission was appointed to investigate the accident.

Soon, it became clear that many missions before STS-51L could have resulted in a *Challenger*-like accident. Rubber O-rings in the seals between the joints on the solid rocket boosters were susceptible, especially to damage in cold temperatures. Some had leaked hot gases before, but the O-rings' seals

Below: The Space Shuttle *Challenger* lifts off from Pad 39B at the Kennedy Space Center on January 28, 1986.

had never been breached completely, as had happened in the case of STS-51L.

The breach, at SRB ignition, mysteriously sealed itself and opened again about 50 seconds after launch, causing the attach-ring to collapse, the booster to crash into the external tank, and the vehicle to break apart.

The Rogers Commission made a damning report of oversights and NASA's refusal to accept warnings from SRB contractor engineers. The SRB seals were redesigned and the Shuttle was ready to fly again in late 1988.

However, observers watching the launch from Smyrna Beach, north of the Kennedy Space Center

saw a third contrail coming from the errant SRB during all of the ascent, as if one of the boosters had another exhaust outlet. Video coverage shows the booster also shedding some material. The Shuttle was fish-tailing, indicated by the in-and-out of focus video coverage of the vehicle, possibly due to a loss of some thrust on one booster, and not by wind shear as NASA had claimed. This was compensated by the gimbaling of the main engines, which can also be seen clearly.

Photographs never published show fiery exhaust

vapors coming out of the breach during the roll program, well after the time it was said to have sealed itself. It appears that the fault may not necessarily have been an O-ring but a breach of the booster casing due to structural failure around the attach-ring area, which seems to be confirmed from the damaged recovered part of the breached booster.

Stresses at launch

When the Shuttle stack lurches forward at main engine ignition, this puts enormous strain on the base of the vehicle, especially around the attach-ring area. The Shuttle then bounces back and with the engines at 100%, the SRBs light. The "twang" of the Shuttle is said to be caused by the dynamic overshoot that resulted in the mysterious damage to

the launch pad and the Shuttle and its payloads that occurred many times during the early pre-*Challenger* missions.

The original scenario had the engines firing for three seconds before lift-off, and the Shuttle was designed for this, but later it was decided to increase the time to six seconds The resulting excessive twang and the dynamic loads were causing the damage.

NASA redesigned the Space Shuttle O-rings but also did many other things, including strengthening

the half-clamp that connects the SRBs to the orbiter—the point of the greatest strain on the system during the twang—with a full circumferential clamp. This modification has hardly been mentioned by NASA PR or the media.

This and other strengthening of the system added 22,050 lb (10,000 kg) to the SRB's weight and therefore an equal reduction in the payload carrying capacity of the system. Whatever the cause, one thing is certain. The Shuttle was considered to be a safer vehicle, was regarded in a different light, and with much more respect. One former astronaut commander said that the Shuttle was in a test phase for the first 100 missions.

Center: The crew of *Challenger* (left to right): Ellison Onizuka, Mike Smith, Christa McAuliffe, Dick Scobee, Greg Jarvis, Judith Resnik, and Ron McNair.

Below: The Space Shuttle *Challenger* breaks apart in an explosion at T+73s into the launch. The Space Shuttle fleet was grounded for 21 months.

The Operational Shuttle

Space Shuttle operations resumed in September 1988, with STS-26 spearheading the return to space. A line was drawn—there would be no more gimmicks, just functional and realistic operations.

Space servicing was featured by flights to the Hubble Space Telescope, with the Shuttle likened to a flying garage. There were also missions to the Russian Mir space station, and later regular flights to begin construction of the International Space Station (ISS).

Most missions were unheralded and featured science aboard Spacelab, satellite deployments and retrievals, R&D flights, and Department of Defense missions. A military flight was also launched in 1988 and was followed in 1989 by five missions, including the deployment of the Magellan Venus and Galileo Jupiter orbiters.

Right: The Hubble Space Telescope receives a service from spacewalking Shuttle astronauts.

Below: A routine Space Shuttle mission ends, with the orbiter landing on the runway at the Kennedy Space Center with the aid of a drag chute.

The year 1990 featured a five-month grounding of the whole Shuttle fleet after dangerous fuel leaks, but there were six missions, including the deployment of the Hubble Space Telescope, the first in a series of four planned NASA Great Observatories. The last 1990 mission was led by Vance Brand, the first space traveler over 60.

A decade of successes

There were six flights in 1991, including the deployment of NASA's second Great Observatory, the Gamma Ray Telescope, and an intensive Spacelab Life Sciences mission with two payload specialists, a medical doctor and a vet. A military mission carried a military space reconnaissance specialist.

The eight missions of 1992 starred the maiden flight of *Challenger*'s replacement, *Endeavor*, on an extraordinary mission to capture the stranded Intelsat VI communications satellite, fix a new rocket motor to it, and re-deploy it back into orbit. After the first capture attempt failed, a unique three-person EVA was made to capture the satellite by hand, and it eventually made its way into geostationary orbit.

The seven flights of 1993 demonstrated the routineness and versatility of Shuttle operations with satellite deployments, an Atlas science mission, the introduction of the Shuttle extension module called Spacehab, a life sciences mission, and the first Hubble Space Telescope servicing mission.

Seven 1994 missions included the first Russian to fly on an American mission, Sergei Krikalev, in preparation for the Shuttle Mir Mission (SMM) program, and the introduction of the Space Radar Laboratory for versatile, high-resolution Earth topography observations.

1995 witnessed the first flight to Mir, to pick up American Norman Thagard, who had earlier been launched on a Russian Soyuz rocket. It was also the 100th U.S. manned spaceflight.

There were seven flights in 1995 and a further seven in 1996, including more missions to Mir, a reflight of the Italian tethered satellite which was lost when the tether broke, and the flight of Story Musgrave on a then unique sixth Shuttle mission.

The eight missions in 1997 featured an aborted Spacelab microgravity mission, which came home after three days due to a fuel cell fault, but which was launched again 88 days later with the same crew—another historical "first."

The year 1998 saw five missions, including the epic flight of a certain 77-year-old—John Glenn, who was the first American in orbit, in 1962 *(see page 38)*. This was justified as a medical mission on ageing but was really just a well-deserved second flight by an American hero, who, it had been revealed, was grounded by President Kennedy as Glenn had become an American icon. The year ended with the first Shuttle mission to the ISS.

The first Shuttle flight in 1999 did not take place until May, with another mission to the ISS. Other missions included the deployment of the third Great Observatory, the Advanced X-Ray Facility, and only one other mission was flown, in December. There were delays due to technical problems with the fleet.

The year 2000 saw the 100th Space Shuttle mission, and the vehicle could at last (in test-piloting terms) be called operational. In 2000/2001 eleven missions were dominated by ISS work and crew transfers, with two astronauts making a seventh flight. There was also another Hubble service and two ISS missions, before delays due to engine problems with the fleet.

The Future of the Shuttle

There is little likelihood of a second-generation reusable launch vehicle (RLV) flying before 2010. The step from the Space Shuttle to an RLV like an airliner is huge, both technically and financially. However, with continuous upgrades and improvements being made, the present Shuttle fleet could be flown until 2010 or after.

Right: Liquid propellant-powered fly-back boosters could replace the current solid rocket boosters, providing improved safety, control, and reusability.

Now that the Space Shuttle has flown over 100 missions, it is considered to have passed the test plane stage and become operational. All the same, flying on a Shuttle mission is still a risky business, and—as February 1, 2003 proved—an accident is always possible.

Although the Space Shuttle may look like the same vehicle that was designed in the 1970s, many critical improvements have been made to it and will continue to be added. One major innovation is the so called glass cockpit. During the dynamic launch of the Space Shuttle, milliseconds make a lot of difference. Glass cockpits enable a crew to react faster and be better able to avoid a disaster.

The first mission to fly with the glass cockpit was an International Space Station flight, STS-101/*Atlantis*, in May 2000, with commander James Halsell and pilot Scott Horowitz at the controls. The new Boeing 777-style glass cockpit will eventually be installed on all the Space Shuttle orbiters.

Dozens of outdated electromechanical cockpit displays like cathode ray tube screens, gauges, and instruments will give way to full-color flat panel screens. These provide Shuttle crews with easy access to vital information through the two- and three-dimensional color graphic and video capabilities of its onboard information management system. Not only does the new system improve crew/orbiter interaction with the easy-to-read, graphic portrayals of key flight indicators like attitude-display and Mach-speed, but it also reduces the high cost of maintaining obsolete systems.

The hardware consists of 11 identical full-color liquid crystal Multifunction Display Units (MDU). Four of these directly replace the four monochrome units of the old system; two MDU's replace the commander (CDR) and two more the pilot (PLT) flight instruments; one MDU replaces the in-orbit maneuvering instruments at the aft flight deck; and the remaining two MDU's replace the CDR and PLT status displays. The command and data entry keyboards, as well as the rotational and translational hand controllers, and most of the other cockpit switches, remain unchanged.

Other Shuttle improvements will be made over the years, as budgets allow. These include electric auxiliary power units to replace the present hydrazine-powered units, which are extremely expensive to maintain. The original oxygen-hydrogen fuel cells used to generate electricity may be replaced by more powerful proton-exchange membrane fuel cells, while a Space Shuttle Main Engine Advanced Health Management system may be introduced to increase safety and reduce turnaround costs.

Another upgrade planned is a switch of the Shuttle's main propulsion system propellant valve from pneumatic to electromechanical actuation. Others include more durable thermal protection system tiles for the underside of the Orbiter and changes to the main landing gear tires, and improved abort systems.

Learning from the Columbia disaster

Not every improvement may seem to be important, but all together, the advanced Shuttle systems will enable it to fly for many more years to come. One improvement that will make the Shuttle look very different is the proposed liquid fly-back booster. The twin solid rocket boosters (SRB) of the Shuttle are effective but do not offer much of a safety margin, nor the most versatile flight control options. Though not fully funded yet, the SRBs may be replaced with ones which are propelled with liquid oxygen-liquid hydrogen engines, similar to those that are carried on the Orbiter.

These liquid fly-back boosters (LFBB) would be equipped with wings, and when jettisoned, would be able to fly back automatically to the Kennedy Space Center, landing like airplanes on

Right: The Space Shuttle's "glass cockpit" will eventually be introduced to all the orbiters.

the runway. The LFBBs would be safer, more reliable, less expensive, and would have a better performance that the old SRBs.

The accident during re-entry which killed the seven crew of the Space Shuttle *Columbia* on February 1, 2003 should not be the kind of body-blow that hit NASA after *Challenger*'s loss in 1986. The program will undoubtedly be grounded, probably until late 2003 or 2004, but the accident served to remind everyone that exploring and using space is dangerous and always will be. The best tribute that can be paid to the lost STS 107 crew will be to continue to fly as soon as possible and safely as possible, but also to accept that the risk is worth it. If not, we might as well shut the door on human space exploration.

Death of the Soviet Dream

By 1988, the Soviet Union had developed a new super-booster called Energia and a Space Shuttle vehicle called Buran, and planned to operate a second generation Mir space station. Mir 2 would comprise of modules flown up into orbit on the Energia or in the payload bay of Buran, to establish a permanently manned base in space that would be a stepping stone to Mars. It was a logical scheme, on the same lines as American plans. In fact, Buran was almost a direct copy of the American orbiter.

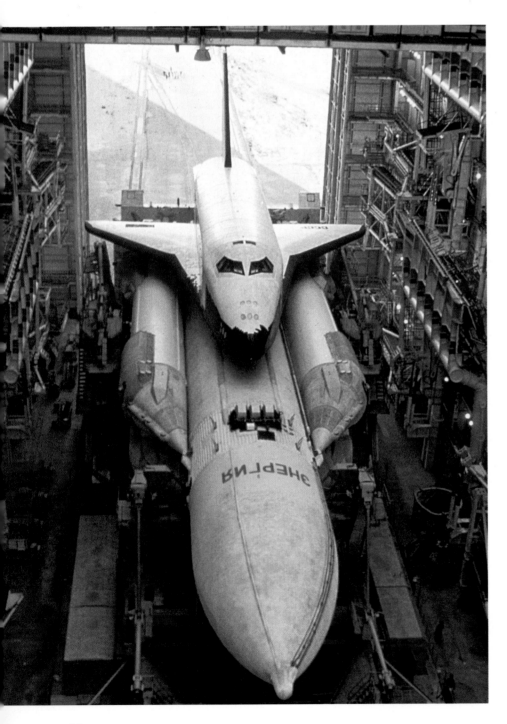

The Soviet government gave the Energia-Buran project the go-ahead in 1976. Energia was an all-liquid propellant vehicle on which the Buran orbiter would ride piggyback style, just like the U.S. Space Shuttle, although the Shuttle flies on an external tank which provides propellant for its own engines and two huge solid rocket boosters.

In 1977, the same year that NASA was conducting glide tests using the piloted *Enterprise* prototype orbiter, the Soviets tested a prototype, called Spiral, dropped from a modified Tupolev-95 strategic bomber for a free-flight test. In December 1980, a Bor spaceplane conducted its first sub-orbital flight on a Cosmos booster to test Buran's thermal-protection system, and in June 1982—when the fourth U.S. Shuttle mission was flying—a Bor was launched into orbit and splashed down in the Indian Ocean. To cover up the test, the flight was called Cosmos 1374.

A model of the Buran shuttle orbiter was carried aboard a Soviet carrier aircraft in March 1983, while a Bor was launched into orbit as Cosmos 1445 and splashed down in the Indian Ocean. An indication that the Soviet Union was testing a Shuttle was given when the recovery of the craft by a Soviet Navy vessel was photographed by an Australian military plane, and the pictures published around the world.

In December 1983 another Bor spaceplane was launched into orbit as Cosmos 1517, and splashed down in the Black Sea instead of the Indian Ocean, to maintain secrecy around the project.

The first launch of a Bor 5 with a full-size model Buran orbiter was made from Kapustin Yar, on a sub-orbital flight, in July 1984, but the model didn't separate from the booster. After a second Bor 5 flight, a prototype of Buran conducted the first atmospheric flight, with Igor Volk and Rimantas Stankavichus, in November 1985. The Bor 5 model made three flights in 1986–88.

The first automated approach and landing tests by the Buran prototype, and an entirely automatic flight, were made in February 1987. On May 15, 1987, the Energia booster—the most powerful rocket in the world—was launched from Baikonur, carrying a Polyus military payload. The rocket performed flawlessly but the Polyus orbital maneuvering system fired in the opposite direction,

The Soviet Union's Space Shuttle system comprised of the Energia Heavy Lift Booster and a spaceplane based on America's Space Shuttle. The first and only spaceplane to fly was a Buran and made just one unmanned flight from and back to Baikonur in November 1987, before the program was canceled.

due to a control system problem, causing the payload to fall into the ocean.

The Energia, carrying an unmanned Buran reusable shuttle, blasted off from Baikonur on November 15, 1987, with the winds 5 mph over launch criteria. The launch went well, however, and 206 minutes and two orbits later, the Buran automatically landed back at the launch site.

Buran was able to carry a payload of 105 tons, returning with 87 tons. It could be crewed by ten people and was capable of flying for 30 days. Buran was 118 ft (36 m) long, with a wingspan of 78 ft (24 m).

Best-laid plans...

The Energia and Buran programs struggled amid the turmoil of the collapse of the Soviet Union, and on June 30, 1993, Boris Yeltsin canceled them. It was later confirmed that the first flight of the new orbiter, with life support systems and ejection seats, and two cosmonaut test pilots—Volk and Stankavichus—was to have been launched in 1992.

Volk—who had previously flown a familiarization flight aboard a Soyuz craft to the Mir space station—and Stankavichus, would have delivered a module to Mir, using the Buran manipulator arm to dock it to the station's Kristal module. The collapse of the Soviet Union, however, marked the end of these plans.

CHAPTER SEVEN

SPACE STATIONS

Skylab

During 1966, as NASA was making final plans for its assault on the moon, the agency was simultaneously looking to the future. Its goal was a permanently manned space station in Earth orbit, as a stepping stone to Mars. As an initial step it was decided to build an interim space station, using components developed for Apollo.

The project was called the Apollo Applications Program, or AAP, which would be launched during the final Apollo flights to the moon. AAP became known as Skylab. It would comprise the empty S4B third stage of its Saturn 5 launcher, and astronaut teams would be flown to it aboard Saturn 1B boosters in Apollo command and service modules (CSM), and would convert the stage into a working station. The use of the Saturn 5 for Skylab reduced Apollo by one flight.

Then the plan changed. The S4B stage would be a fully equipped station when it was launched. Skylab was also equipped with a telescope based on a lunar module structure.

The main part of Skylab was the Orbital Workshop (OWS), the S4B structure, 48 ft (14.6 m) long and 22 ft (6.7 m) in diameter, providing 10,000 sq ft (283 sq m) of living and working space on two levels divided by a grid floor. The lower level contained the wardroom, kitchen, washroom, and living quarters, and was equipped with a large picture window for viewing the Earth. Medical equipment was also located in the lower level. This had an airlock on the floor leading to the empty liquid hydrogen tank, which was used as the trash can. The upper level contained most of the crews' consumables and work areas.

The top of the OWS led to an airlock module, from where space walks could be made. This airlock led to the Multiple Docking Adaptor (MDA), which could enable more than one CSM to join with the station, for example in the event of a space rescue being required in the event of a failed CSM.

Solar observation

Attached to the side of the MDA was the most important instrument on Skylab, the Apollo Telescope Mount. this was equipped with five telescopes observing in different wavebands, mainly to study the sun. Indeed, one of Skylab's major accomplishments was the mass of data collected about the Earth's nearest star.

Skylab was launched on May 14, 1973—two years after the Soviet Union's Salyut 1. The first crew was scheduled to follow Skylab into orbit a day after launch. The launch was not totally successful, however, and at first the space station

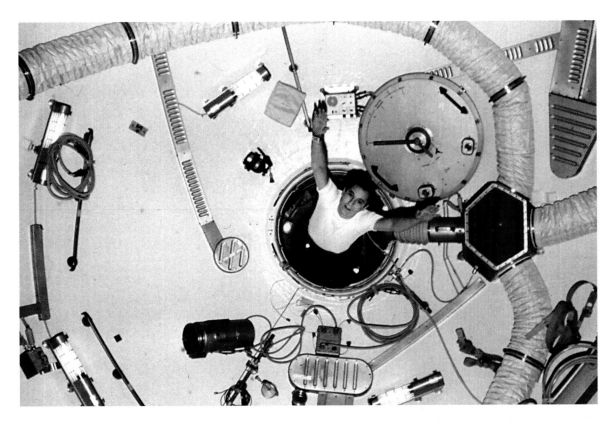

Opposite: America's first space station, *Skylab*, was based around the third stage of a Saturn 5 booster, with an Apollo Telescope Mount designed around a Lunar Module framework.

Left: Edward Gibson floats through a hatchway on the Skylab station during the then record-breaking 84-day flight in 1973–74.

Below: Jack Lousma making a spacewalk from Skylab during the second mission in 1973.

looked doomed. A meteoroid shield on the side of Skylab was torn off during launch, taking with it one of the space station's two stowed solar panels and jamming the other. The first manned mission, dubbed Skylab 2, was now to be a full-scale rescue rather than a routine flight. Veteran moonwalker Pete Conrad took his rookie crew of two to Skylab and, over 28 days—during which a bold and dangerous spacewalk was carried out to extract the remaining solar panel—the station was converted into a workable space base.

The 59-day Skylab 3 mission in 1973, led by Al Bean and regarded as the most successful, was followed by Skylab 4's record to 84 days in 1973–74, set by Gerry Carr, Edward Gibson, and Bill Pogue. This crew became rather infamous for being rebellious with mission control.

When they left, the station was abandoned. It could have lasted many more years, but there were not enough Apollo rockets left to fly astronauts to the station. Budget cuts were hitting hard. A crew to the next U.S. space station would not be launched until 1998.

Skylab was a great scientific achievement, and it is sad that it was left empty in orbit for five more years before making a spectacular re-entry, showering parts of the Australian outback with small pieces of debris.

Salyut

Despite the Soviet Union's failure to put men on the moon, throughout the 1970s and early 1980s the nation dominated the manned space arena, with a series of highly successful space stations. The plan was for long-duration spaceflight experience to be gained in preparation for a planned return trip to Mars.

Above: An artwork of Salyut 1 with the Soyuz ferry attached (left). Salyut 1 was the first space station, launched in 1971, and housed a crew which was killed during re-entry after a three-week mission.

The world's first space station, Salyut 1, was launched on April 19, 1971. It was a simple, single, pressurized module. Salyut 1 weighed 41,600 lb (18,900 kg) and was 47 ft (14.5 m) long. It was equipped with two pairs of solar panels, extended from the rear instrument compartment and from the forward of three workstations respectively.

The docking and transfer section to receive the Soyuz ferry and its crew led to the first workstation and a large work compartment, 13 ft (4 m) long and wide. At the rear was the propulsion system, which housed the maneuvering engine. Salyut carried telescopes, cameras, and sensors for astronomy, science, and Earth observation, as well as exercise equipment. Salyut 1 re-entered the Earth's atmosphere in October 1971, after housing just one crew for 21 days.

Salyut 2 was launched in 1972, and although details of the station have never been released, it appears that this resembled Salyut 1. The workstations were basically of the same design, but at the front was a conical capsule which was designed to undock and return to Earth. Salyut 2 was probably a military space station, equipped with a 32 ft (10 m) focal length camera with a resolution of 3 ft (1 m). Film was returned to Earth in the capsule.

Salyut 2 actually failed in orbit and was never manned, but the similar Salyut 3 and Salyut 5 stations followed in June 1974 and 1976 respectively. These were manned by three two-person crews. Salyut 3 re-entered the Earth's atmosphere in January 1975 and Salyut 5 in August 1977.

Multi-purpose laboratories

In between these military missions, the civilian Salyut 4 was launched in December 1974. The body of the space station was almost identical to Salyut 1, but instead of two pairs of solar panels, Salyut 4 had one set of three. One of the major

instruments on Salyut 4 was the Orbital Solar Telescope (OST), which took up most of the space of the rear section of the larger workstation. The OST was one of seven astronomical experiments on the station, which were complemented by medical and biological experiments. Two long duration crews worked aboard this Salyut, which re-entered the Earth's atmosphere in February 1977.

It was followed by Salyut 6, launched in September the same year, the first of a second generation series of Salyuts. The first unmanned Soyuz-class ferry type, called Progress, was launched and docked to the rear of the station to provide fuel, water, oxygen, food, mail, and personal possessions for the crew. Salyut 6's interior was dominated by a large telescope with a liquid nitrogen cryogenic cooling unit. Other instruments included stereographic, topographical, and Earth resources cameras.

Regimented work aboard Salyut 6 often involved routine maintenance spacewalks. A new

module, called *Star*, docked with Salyut 6 in April 1981. Designated Cosmos 1267, it was used to perform the de-orbit burn for Salyut 6 in July 1982 because the station's own propulsion unit was not working correctly.

Salyut 7 was launched in April 1982 and was joined by two Salyut-class Heavy Cosmos modules, 1443 and 1686, adding more workspace and solar power, while 1443 also released a re-entry capsule during its time docked to the station.

Toward the end of its life, Salyut 7 became severely degraded and eventually contact was lost with the station in February 1985. Two cosmonauts were launched in June that year to restore the station to life, after which Cosmos 1686 was launched to support one more manned mission and a short visit from a crew from the new Mir space station in 1986. Salyut 7 was abandoned and made a natural re-entry in February 1991. The Soyuz missions to the Salyuts were like a regular taxi service.

Below: Salyut 7 prepares to receive a visit from a Heavy Cosmos module.

The Space Taxi

If an American Apollo command and service module was still flying today, it would be a good comparison with the Russian Soyuz. The craft is the longest serving manned spacecraft in the world, having flown almost 100 manned missions from 1967 to 2003.

Below left: A Soyuz U booster launches a Soyuz TM spacecraft carrying three crew from the Baikonur Cosmodrome.

Below right: The Soyuz spacecraft is inserted into the payload shroud of its launcher before roll-out to the pad. Pictured is the forward orbital module and docking port and the flight and descent capsule. The propulsion module cannot be seen.

The Soyuz was designed in the early 1960s as part of Sergei Korolov's bid for initial human exploration of the moon, codenamed OKB-1. The Soyuz was to be a three-part spacecraft assembled in low-Earth orbit, but gradually the OKB-1 plan was downscaled. For the manned lunar landing bid, Korolov's plan was very similar to Apollo, with a two-man Soyuz version being planned, in contrast to the later N-1 booster-launch plan that was eventually approved. A Soyuz version was renamed Zond, and attempted flights that could have led to a manned circumlunar mission before Apollo 8. However, this did not materialize (*see page 80*).

Korolov also originally designed a Soyuz version called the 7K-OK, for missions in Earth orbit for practicing the rendezvous maneuvers for a lunar mission. This was very similar to the American Gemini. Realizing that America was going to win the moon race, Korolov diverted his efforts to space stations, and the Soyuz became the space taxi, an upgraded version of which is still flying today to and from the International Space Station.

In 1967, however, in the midst of the moon race, the Soviet Union planned to launch a space spectacular. Two manned Soyuz spacecraft were to perform a crew exchange in orbit. This ended in disaster with the death of a cosmonaut.

The Soyuz was capable of flying three crew but without space suits, a dangerous plan which backfired in 1971, when the first crew of the Soviet Union's Salyut 1 space station were killed when the craft depressurized. A redesigned, two-seat Soyuz spacecraft—with two spacesuited cosmonauts—then continued ferrying the crews to Salyut space stations.

There were two versions, one the standard module, and the other a battery-powered version with no solar panels, limited to just a two-day flight to dock to Salyut space stations. If a docking did not take place, the craft had to return immediately to Earth, as happened a number of times.

Also, in 1975, a specially-modified Soyuz was developed for the joint U.S.-Soviet Apollo-Soyuz program, in which the two space rivals joined, with three Americans in an Apollo, equipped with a

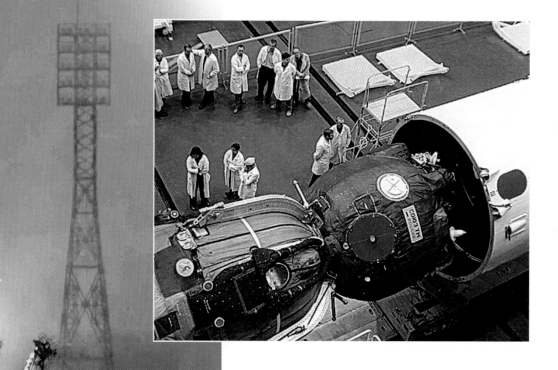

docking module joining to a two-man Soyuz, using an androgynous docking mechanism.

Unmanned Progress versions were used as disposable space tankers for space station support, and introduced in 1978. The upgraded Progress M is now flying ISS missions.

Increased sophistication

Soyuz T and TM versions were introduced in 1980 and 1986, with the three-crew capability restored. The Soyuz TM modification was used for the Mir space station program. The modifications featured multiple improvements in the design, including the introduction of the new weight-saving computerized flight-control system and improved emergency escape system.

The Soyuz TM weighs 15,640 lb (7,100 kg) and comprises a 2,860 lb (1,300 kg) orbital module with a docking mechanism at its front, a 6,390 lb (2,900 kg) flight crew cabin or descent module, and an instrument and propulsion module weighing 5,730 lb (2,600 kg). It can carry 66 lb (30 kg) of payload up to the ISS and 110 lb (50 kg) down. The maximum length of the spacecraft is 22 ft (7 m) and the maximum diameter, 8.9 ft (2.72 m). The descent module has a diameter of 7 ft (2.2 m), and the craft has a twin-solar array span of 35 ft (10.7 m). The craft, descending on a single main parachute, fires retro rockets just before landing, to cushion the impact further at a speed of 6.6 mph (10.6 km/h)

A new version, called the Soyuz TMA, removed the limitations for the height of the crewmembers, allowing the craft to be used as a lifeboat for the ISS, and was flown for the first time in October 2002.

Above: A Soyuz TM spacecraft prepares to dock with the International Space Station.

Russian Routine

Soyuz began as a rushed and botched attempt to match the successes of Apollo, and many cosmonauts were killed during the early missions as a result of the program's undue haste. As the series progressed, however, great strides were made, and by the mid-1970s the Soviet Union had by far the most experience of long duration spaceflight in Earth orbit. It is a lead that Russia maintains to the present day.

Below: Vladimir Komarov trains for his ill-fated Soyuz 1 flight. The spacecraft was clearly not ready to be manned and Komarov was flown for political reasons which backfired.

The first Soyuz spacecraft had not been fully tested when on April 23, 1967, Vladimir Komarov was launched into orbit with the plan that a day later Soyuz 2, with three crew, would dock with him and transfer two crewmen during a spacewalk. Soyuz 1 suffered many malfunctions, and the launch of Soyuz 2 was canceled. After struggling in orbit, Komarov finally made it back, only for his craft's parachute to fail. His capsule smashed into the ground and caught fire. The mission was the most irresponsible in manned space history. Pictures of Komarov going to the launch pad reveal his almost resigned acceptance of martyrdom.

Soyuz 3 failed to dock with an unmanned Soyuz 2 in October 1968, but in January 1969 the Soyuz 1 flight plan was completed with a crew transfer and return to Earth, using Soyuz 4 and 5. A similar flight was planned with a Soyuz 6–8 mission involving seven cosmonauts, in October 1969, but

did not materialize, and in June 1970 Soyuz 9 was flown for a record 17 days to test the two-man crew's reaction to long duration spaceflight.

Risky business

Soyuz 10 initiated space ferry operations with an unsuccessful docking attempt at Salyut 1, the world's first space station, in April 1971. This was achieved by Soyuz 11 in July, and the three crew stayed a record 23 days in orbit, but the unspacesuited trio died when the craft depressurized during the return to Earth. Soyuz 12 tested the solar panel-less ferry version of the craft in September 1973, but another version failed to

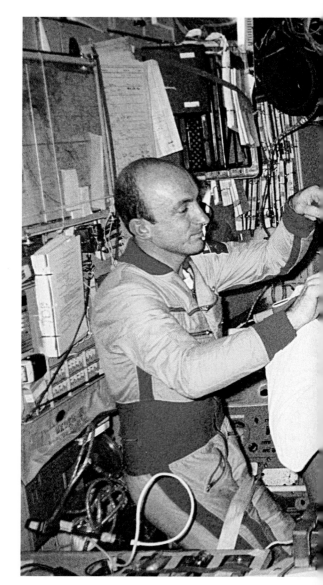

dock with Salyut 3 and came back to Earth immediately in August 1974.

A planned Soyuz 18 launch was aborted at staging in April 1975, while the first in a series of long duration crews on further Salyut civilian and military stations were launched, several featuring docking failures and emergency returns to Earth. One of these almost resulted in the loss of a crew.

The Apollo-Soyuz mission in July 1975 involved Soyuz 19, crewed by Alexei Leonov and Valeri Kubasov, being joined by Apollo 18's Tom Stafford, Deke Slayton, and Vance Brand. It was a great example of détente, but a similar mission was not repeated for 19 years. Apollo-Soyuz provided the U.S. with a manned space presence while the Space Shuttle was being developed.

The 1975 Salyut 4 space record of 62 days was

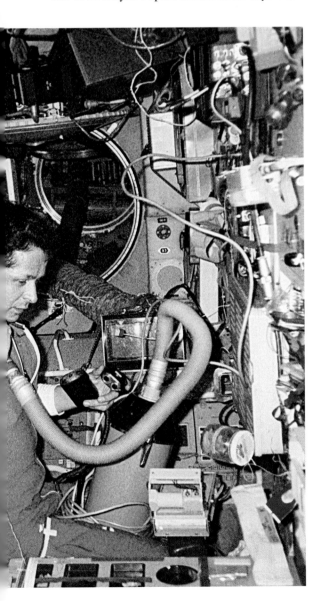

exceeded by 30 days by the Soyuz 26/Salyut 6 crew in 1977, while the Soyuz 29, 32, and 35 crews flew missions lasting 139, 175, and 184 days in 1978–80, the latter two missions being flown by Valeri Ryumin.

The Soviet Union was amassing a wealth of data on long-duration spaceflight as well as conducting a range of experiments, especially in materials processing and also Earth observation and astrophysics.

The nation was also enhancing its relationship with fellow communist states under an Intersputnik program, which involved short flights to space stations. The first space passenger was a Czechoslovakian cosmonaut, Vladimir Remek, in 1978. Others included pilots from Mongolia and Vietnam.

The space endurance record broke the 200-day mark in 1982 with the flight of the Soyuz T5 cosmonauts, Anatoli Berezevoi and Valetin Lebedev, aboard Salyut 7. A French cosmonaut flew the same year.

A crew was saved by an emergency escape system after a launch pad explosion in 1983, and by 1984 the space duration record had been raised to 236 days by Leonid Kizim, Vladimir Solovyov, and a doctor, Oleg Atkov.

The Soviet manned spaceflight monopoly received a setback with the launch of the first U.S. Space Shuttle, in 1981, but another Soviet space station was waiting in the wings to extend the Russian lead to staggering proportions.

Above: The first joint Soviet-U.S. manned mission in July 1975 saw the docking of Apollo and Soyuz spacecraft. Pictured left to right are Tom Stafford, Alexei Leonov, and Deke Slayton.

Left: Onboard Salyut 7 in 1985, cosmonaut Vladimir Vasyutin cuts Viktor Savinykh 's hair.

Magnificent Mir

When the Mir space station was jettisoned into the atmosphere to burn up in 2001, to make way for the International Space Station, one of the greatest achievements of the Space Age came to an end. Mir had its problems, but the very fact that it overcame them is an illustration of Russia's advanced capability, which included supporting a record breaking 437-day spaceflight by Valeri Poliakov.

Mir got itself a bad name mainly because the technically-sophisticated West got involved in the project and didn't like the functional Mir, which experienced breakdowns, a fire, and a collision. The West was applying double-standards, however. When assembly of the International Space Station began in 1998, not all the U.S. component parts worked perfectly either, but they have not yet been likened to behaving like a used car, as Mir was described.

The 46,000lb (20,900kg) Mir core module was launched on February 26, 1986, and included a multiple docking port to receive up to five vehicles. A Kvant 1 module, docked to the rear of Mir on April 9, 1987, weighed 24,360 lb (11,050 kg), and Mir now measured almost 62 ft (19 m) in length.

The 40,790 lb (18,500 kg) Kvant 2 module was added to the front port on December 6, 1989, and was laden with an array of telescopes, cameras, and equipment. *Kristall*, weighing 43,300 lb (19,640 kg) was next to join the complex, on June 10, 1990, with a small robot arm being transferred to an

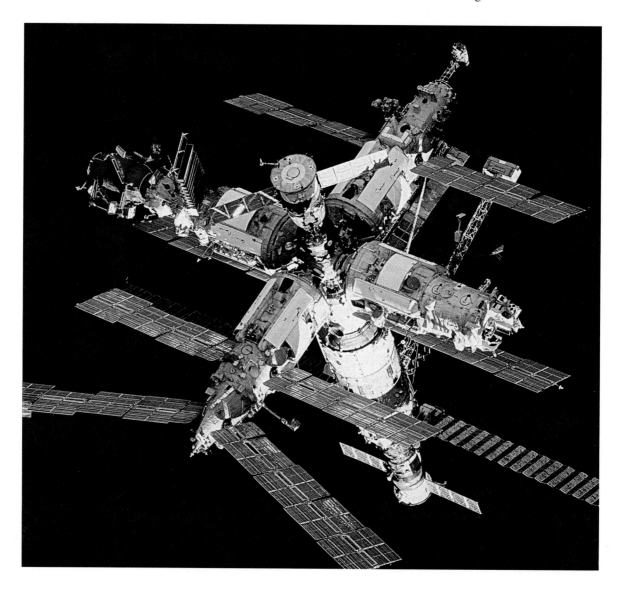

Right: The fully-assembled Mir space station pictured in 1995 from the Space Shuttle, during one of the joint Shuttle Mir Mission flights.

adjoining port the day after. *Kristall* was dedicated mainly to materials processing experiments.

The next module to join was *Spektr* on June 1, 1995. Weighing 43,300 lb (19,640 kg) at launch, it was dedicated to Earth sciences and atmospheric monitoring. The final Mir module, the 43,430 lb (19,700 kg) *Priroda*, arrived April 26, 1996. This was equipped with an array of remote sensing cameras.

Sixteen-year saga

The remarkable Mir enabled cosmonauts to clock up flight times of over a year, amassing a wealth of biomedical data about the effects of long duration spaceflight on the human body. A continual stream of cosmonaut teams were launched to Mir starting in 1986, and were still being launched there in 2000. These teams also included crew persons from many other countries, making Mir the first international space station. Later in the program, after the collapse of the Soviet Union, Russia charged for these foreign trips and for experiment time on the station. Regular and routine EVAs were made outside Mir to conduct experiments and make repairs.

In March 1986, Soyuz T15's Leonid Kizim and Anatoli Solovyov made a unique 125-day mission to both Mir and Salyut 7, and were followed in February 1987 by a crew including Yuri Romanenko, who made a record-breaking 326-day flight.

Several international missions were made to Mir, starting with a Syrian cosmonaut in June 1987, while two of Soyuz TM4's three-man crew, Vladimir Titov and Musa Manarov, made a record breaking 365-day flight.

Further routine missions followed, and in December 1990 Soyuz TM11's crew included Japanese journalist Toyohiro Akiyama, the first commercial passenger in space, making a miserable seven-day flight feeling very sick. True to a journalist's reputation, when he landed he asked for a beer and cigarette!

Starting in May 1991, Soyuz TM12's Sergei Krikalev flew for 311 days, leaving Earth as a citizen of the Soviet Union but returning as a Russian, when the USSR broke up. In March 1992, Soyuz TM14 made the first official Russian spaceflight, and in 1994 Poliakov started his record mission.

In 1995, the U.S.-Russian Shuttle/Mir Mission program began, involving several long duration

missions by U.S. astronauts, and in June 1995 the historic STS-71/*Atlantis* Shuttle/Mir Mission 1 crew docked with Mir.

The Soyuz TM28 three-man crew in August 1998 included Yuri Butarin, a cosmonaut observer and former presidential aide, who flew for 11 days. The main mission lasted 198 days, and as a result Sergei Avdeyev amassed a record of 748 days in space on three missions.

On April 4, 2000, Soyuz TM30's cosmonauts Sergei Zaletin and Alexander Kaleri made the last mission to Mir, lasting 72 days. The 14-year saga of Mir had come to an end, and except for a few weeks, it had been inhabited continuously since 1987.

Above: A Soyuz spacecraft prepares to dock with the Mir space station during an early stage of its assembly.

Below: The busy interior of Mir came as a shock to its American visitors, who found the functional way of life difficult to adjust to.

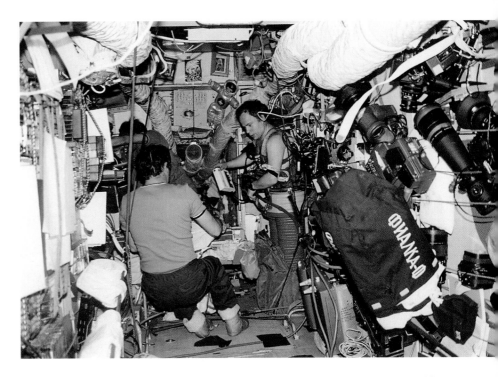

The Quest for Freedom

After Skylab was abandoned in orbit in 1974, America was grounded. The euphoria of Apollo had faded and budgets had been drastically cut. NASA badly wanted a space station, and designed the Space Shuttle to support it. In a contradictory move, President Nixon refused the space station but gave the Shuttle the go-ahead. It signaled the beginning of a confused era in the U.S. space program.

Below: President Ronald Reagan and his wife greet the crew of the returning Space Shuttle STS-4 in June 1982. Reagan was under pressure to announce America's next goal in space.

The planned Shuttle had nowhere to shuttle to and back from, so it was earmarked as a space truck that was to serve government and civilian needs, flying once a week, like a cargo plane. However, the Soviet Union's continued human presence in space and its experience in long-duration flight aboard Salyuts, and especially its announced aims for a mission to Mars, continued to rankle America.

With the Cold War still in full swing, the political climate was similar to that in the early 1960s, when the Space Race was triggered. To counter the threat of Soviet missiles in the early 1980s, President Ronald Reagan introduced Star Wars, the Strategic Defense Initiative, (SDI) which later became known as the Ballistic Missile Defense Organization (BMDO). The Soviet Union was still seen as the Evil Empire.

NASA, having got the Shuttle flying at last after a three-year delay, and realizing that it was not going to fly 50 times a year but ten at the most, was pushing hard for an American space station to counter the threat, and give the U.S. a purpose in space. Reagan somewhat reluctantly gave a space station the go-ahead in 1984, saying that it would be fully operational by 1994. However, he insisted that it must be an international affair, involving Canada, Europe, and Japan.

The budget was set at $8 billion, and NASA wondered how it was going to handle such a project with such an array of partners. Eventually known as *Freedom*, the station's design was extraordinarily ambitious: a huge dual-keel structure which would be constructed by spacewalking astronauts on dozens of Shuttle flights a year, starting in 1992.

Government inaction

Gradually, it dawned on NASA and the U.S. Congress that *Freedom* was far too ambitious and was not going to be built within the budget. It was also becoming behind schedule. The *Freedom* design was down-scaled as a result of the realization of the technical and logistical challenges.

From this point onward Congress complained regularly about the billions being spent on a project that was becoming further and further behind schedule. The development and delivery of hardware was delayed, while the other international partners, such as the European Space Agency, were becoming increasingly frustrated.

Eventually, by 1992, and already $25 billion in the red, with no components in space and with no launches in immediate prospect, Congress told NASA that there would be no *Freedom* —unless Russia joined the project. In 1993, the new NASA administrator, Dan Goldin, brought the Russian and U.S. efforts closer by announcing an interim program—as a way at least of having an international space station. It was called the Shuttle/Mir Mission program, in which crews from both nations would fly together (*see previous page*).

This amazing about-turn came as a result of the collapse of the Soviet Union. The Evil Empire was gone, and its plans for a Mir 2 had been scrapped due to lack of finances. So, both countries could now gain access to each other's space technology, and both wanted a space station. Russia joined the project, which had now become known as the

International Space Station (ISS), while NASA's frustrated original partners were sidelined.

The concept had excellent intentions. Russia, however, was not familiar with Western ways, and what was left of its space program was becoming increasingly starved of budgets. U.S. companies streamed into Russia to bail out space companies, taking the best technology back to America. While Russian space executives enjoyed the attention,

Russian space companies struggled to survive.

Cultural difficulties made relations awkward, and progress slow and complicated. Russia seemed reluctant to fly Americans to Mir, and Americans felt the same way about flying Russians on the Shuttle. An ISS could only be made through such co-operation, though, since the project was too large and costly even for the United States to attempt alone.

Below: The NASA dream of a space station by 1984, with free flying co-orbiting microgravity platforms, experiments mounted on a vast framework, and including solar dynamics mirrors for electrical power generation.

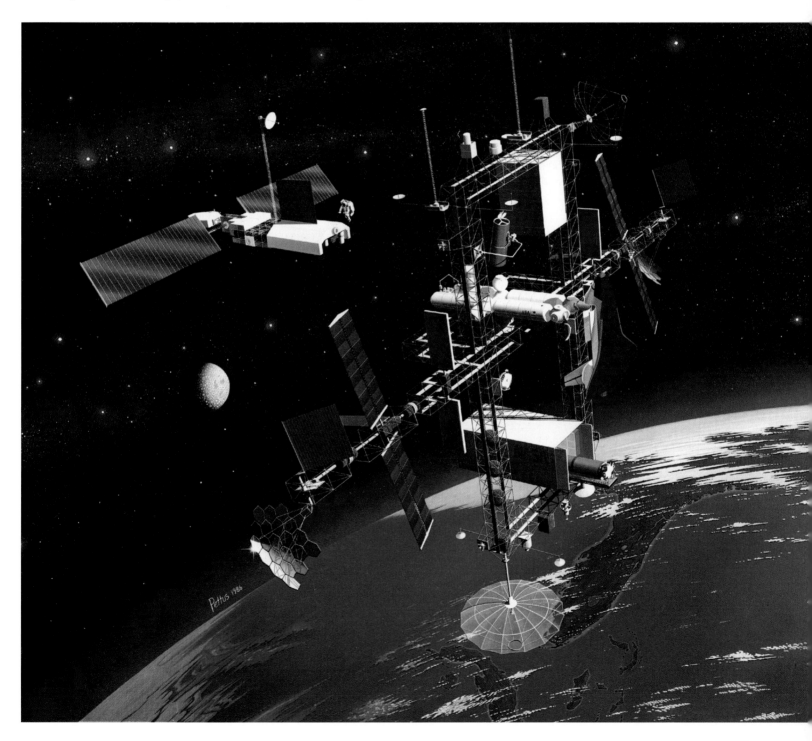

Preparing for the ISS

Two veteran Russian cosmonauts, Vladimir Titov and Sergei Krikalev, arrived at NASA in Houston, Texas, in 1993, to train for a Space Shuttle mission that would spearhead an era of co-operation. Flights of U.S. astronauts to Mir would follow, and work would later begin on the International Space Station.

Krikalev flew STS-60/*Discovery* in February 1994, an independent mission lasting eight days which must have seemed easy for this Mir veteran. Titov flew STS-63/*Discovery* in February 1995 in a rendezvous demonstration flight with Mir, but with no docking, in preparation for joint Shuttle-Mir missions.

On March 14, 1995, history was made with the launch of American astronaut Norman Thagard on Russian Soyuz TM31, on a mission to Mir, where Thagard and his crewmates lived for 115 days.

The historic STS-71/*Atlantis* Shuttle/Mir Mission (SMM) 1 was launched on June 27, and docked to Mir. A new Russian two-man crew, including veteran cosmonaut Air Force Colonel Anatoli Solovyov—a rather reluctant participant in an American mission—was transferred, and Thagard and his crewmates returned to Earth.

Solovyov was not the only Russian to mourn the loss of the once great Soviet space program.

Many of his colleagues resented the American interference. Once they realized that U.S. co-operation was the only way to save what they had left of their space progam, a sense of reality set in.

On the other side, Thagard, a medical doctor and Shuttle mission veteran, found it hard to work on Mir and with his colleagues, who tended to keep him at bay. The operational and cultural differences were the most difficult.

Meanwhile, new Russian Mir crews were to be rotated by standard Soyuz TM switchovers. In November, STS-74/*Atlantis* flew SMM 2, with supplies and equipment, and in March 1996, STS-76/*Atlantis*/SMM 3 delivered NASA's astro-scientist Shannon Lucid for her 188-day flight. Lucid enjoyed her Mir trip and got along well with her colleagues.

STS-79/*Atlantis* flew SMM 4 in September to deliver Air Force pilot and former Shuttle commander John Blaha, and return Lucid. His 128 days were not so easy. By his own admission, he became depressed and found it hard to motivate himself.

Life-threatening emergencies

Another medical doctor, the U.S. Navy's Shuttle veteran Jerry Linenger, flew STS-81/*Atlantis* SMM 5 in January 1997, and was involved in a real emergency—with the Russians and a German visitor—when an oxygen candle fire broke out. His medical experience proved useful, and at the time, an emergency evacuation from the smoky Mir seemed a possibility.

NASA's British-born astronaut Michael Foale was launched on SMM 6 on STS-84/*Atlantis* in May 1997, and impressed the Russians the most, integrating well with the crew. His performance during the most serious in-flight emergency on Mir was also appreciated. While trying to automatically maneuver a Progress tanker in to dock with the station, Mir's commander Vasili Tsiblyev was under pressure because of bulky equipment and lack of experience at this maneuver. The Progress collided with Mir, and the station started to depressurize. Order was restored, but Mir had received extensive damage, and the depressurizing module was quickly sealed—containing most of Foale's possessions. He and his colleagues came close to death.

Below: Shannon Lucid exercises on the Mir space station during her record 188-day mission.

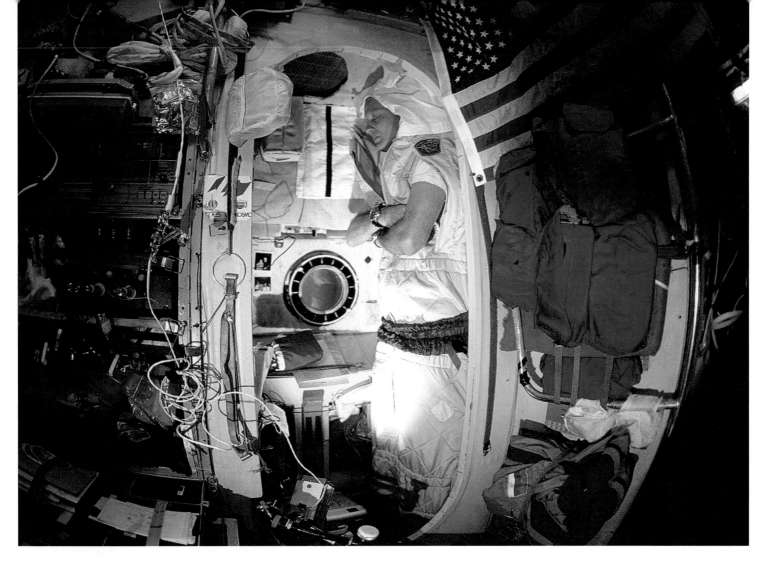

In August, Soyuz TM26, with two crew—led by the no-nonsense Solovyov—was launched on a 197-day mission to make urgent repairs to Mir, and in the following month, STS-88/*Atlantis* flew SMM 7 to deliver medical doctor David Wolf to replace Foale. During the mission, the first joint U.S./Russian spacewalk was made, by Titov and Scott Parazinsky.

In January 1998 STS-89, this time with the orbiter *Endeavor*, was launched on SMM 8 to deliver Andrew Thomas. Wolf's and Thomas' experience had been made easier due to the increasing experience the countries were having working together, which was after all one of the goals of the program.

Finally, on June 2, STS-91/*Discovery* flew the ninth and final SMM with a crew that included an extremely portly Valeri Ryumin, the veteran Salyut cosmonaut from Russia, who talked his way into a flight as the chief of the Mir program.

SMM was a great success and was helpful in keeping Mir operational. Without the experience of the program, the initial orbital operations of the International Space Station would have been difficult.

Above: The first U.S. resident on Mir was Norman Thagard, who made a 115-day flight. He was launched on a Soyuz and returned on the Shuttle.

Left: Ten people gather aboard Mir during the 7th Shuttle-Mir Mission, STS-86 in 1997. David Wolf was replacing Michael Foale as the U.S. resident.

The International Space Station

The ISS is one of the largest international civil, co-operative programs ever attempted, involving 16 nations—the U.S., Russia, Canada, Japan, Brazil, and the 11 countries that comprise the European Space Agency. It is, however, in danger of becoming a present-day version of the Tower of Babel. Reagan's *Freedom* was to cost $8 billion and was to be operational in 1994. If completed, the ISS will have cost over $40 billion and will not now be operational until 2006. International partners are getting very tetchy.

Right: The Russian *Zarya* module and the U.S. *Unity* Node 1 module are pictured joined together in 1998 as the International Space Station takes shape.

Opposite, left: A Soyuz spacecraft is pictured attached to the Russian *Zvezda* module, leading to *Zarya* and *Unity*.

Opposite, right: In 2002, the ISS is taking shape with its first solar arrays, the Destiny laboratory module, a robot arm, the first truss frame, a docking module, and airlock.

Politics, budgets, technical delays, potential accidents, and the lack of any unique, innovative science with direct spin-off on Earth, will likely dictate its future. Russia and the U.S. experienced several problems and the ISS start date was delayed again and again, until at last in October 1998, the first ISS module, the Russian *Zarya*, was launched.

Further delays seem invevitable. The likelihood is that the ISS will be built but that it will not resemble what was originally designed, and will develop on a step-by-step basis, according to the state of delays and finances. If it is ever completed, the space station will weigh one million pounds (453,600 kg), and will measure 365 ft (111 m) end-to-end, the equivalent of a football field.

The ISS was to have been manned by six crew members, always including at least one U.S. astronaut and one Russian cosmonaut. However, that operational ISS Expedition Crew complement has already been reduced to three due to the cancellation of a U.S. Habitation Module and Crew Return Vehicle.

A major part of the station is the Canadian remote manipulator system of two robot arms, one 55 ft (16.77 m) long, and including a mobile transporter that will travel along a rail the length of the station.

Construction site in space

The ISS will be fully assembled from modules, nodes, truss segments, solar arrays, re-supply tugs, and thermal radiators, providing 46,000 cu ft (1,624 cu m) of pressurized living and working space—the equivalent of the interior of a Boeing 747 jet.

The station was to have also included two U.S. laboratories, a European module, a Japanese module, and two Russian research modules, as well as other modules for providing services. This

The electrical power system is connected with 42,000 ft (12,800 m) of wire. The batteries lined end-to-end measure 2,900 ft (883 m) or half a mile (0.8 km). The ISS will have four windows for observing the Earth, offering an all-around, 360° view inside cupola modules which are interconnected in the station. Fifty-two computers will control ISS systems including orientation, electrical power switching, and solar panel alignment.

By mid-2002, the fifth Expedition Crew was working aboard the ISS, and several more were in training for further shifts. Several Russian Soyuz logistics missions had been made to swap spacecraft. One Soyuz is always docked to ensure the ability for an emergency evacuation for the Expedition Crew.

Unmanned Russian Progress vehicles have flown many missions, providing supplies, water, and oxygen, and the facility to take garbage off the station to be destroyed during the re-entry of the departing craft.

Space Shuttle missions delivering hardware and crews, temporarily enlarging the ISS crew to ten, have been flown, but there have also been several delays, the latest in 2002 due to obsolete ground hardware and repairs required on the Shuttle engines.

number has already been reduced and further changes are likely.

Full assembly will have required 45 launches of rockets, mainly the Space Shuttle. Four photovoltaic modules each with two arrays 112 ft (34.16 m) long and 39 ft (11.89 m) wide will each generate 23 kW of electricity. The total surface area of the arrays is about half an acre—27,000 cu ft (2,500 cu m).

The size of the ISS and the reflective qualities of the arrays make the station an impressive sight in the night sky, especially when it can be seen with a similarly star-like Space Shuttle nearby.

Up and Running

Following the placement of the first part of the ISS in late 1998—the Russian module, *Zarya*—an extraordinary and remarkably successful series of complex missions were carried out by the Space Shuttle. These set out to enlarge the station with new modules, truss crossbeams, and robotic systems.

By October 2002, the fifth expedition crew was working onboard. Thirty EVAs had been conducted by pairs of Shuttle spacewalkers, and nine from the ISS itself by Expedition Crews. Ten unmanned Progress tankers had also been launched in this period to support operations, delivering cargo and departing with trash to be destroyed on re-entry.

Following *Zarya* was the first ISS Space Shuttle mission, in December 1998, which delivered the first node connecting module, called *Unity*, and two pressurized mating adaptors. However, the Russian *Zvezda* service module was then delayed to July 2000, and in between, two Shuttle missions were flown in May 1999 and May 2000 to deliver cargo. *Zvezda* was finally launched, and was followed in September 2000 by a Shuttle logistics and *Zvezda* outfitting mission.

In October another Shuttle was launched, on the most complicated mission so far, to attach an Integrated Truss Structure, pressurized module, communications systems, and attitude control gyros. Finally, on October 31, the first expedition crew was launched on a Soyuz booster. Commanded by NASA's William Shepherd, the crew included Russian's Sergei Krikalev and Yuri Gidzenko.

During their space shift, two Shuttle missions were launched in November 2000 and January 2001, to fix the first solar array to the ISS, and an S-band communications antenna, radiators, and other equipment, and the U.S. Laboratory Module, *Destiny*, which itself had been delayed several months.

In March, the second expedition crew; commander Yuri Usachev, Susan Helms, and James Voss, were launched on a Shuttle mission which also carried a new Italian-built returnable logistics module. The first crew returned to Earth on this flight.

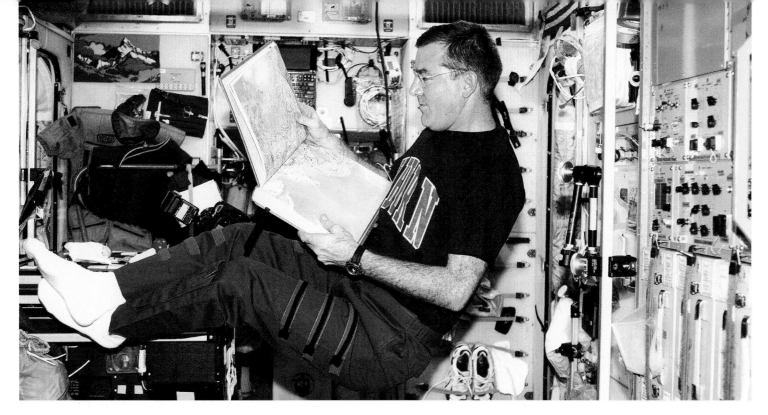

Space tourism begins

Another Shuttle was launched in April 2001 with a logistics module cargo delivery, a UHF antenna, and the first component of the Canadian Space Station Remote Manipulator System. Also in April a Russian Soyuz TM was launched to replace the older Soyuz as an interim Crew Rescue Vehicle (CRV). The mission also carried Dennis Tito, the world's first paying passenger-tourist. Tito paid $20 million for his short visit.

This was followed in July by a Shuttle mission carrying the Joint Airlock and a high pressure gas assembly, and in August, another Shuttle mission was launched with a logistics module delivery and crew transfer.

Expedition crew #3 comprised commander Frank Culbertson with Vladimir Dezhurov and Mikail Tyurin. Another Soyuz TM switch-over, in October, was followed in December by the fourth expedition crew and new equipment. The new crew was commanded by Yuri Onufrienko, with Dan Bursch and Carl Walz as flight engineers.

Alarming cost overruns were revealed in 2001, showing the amount spent on the ISS to that date as $21 billion. This led President George Bush, Jr. to terminate some planned additions, including a CRV spaceplane and habitation module, and to tell NASA to restrict new spending to $26 billion and to complete assembly by 2006.

In 2002, operations continued with a Shuttle mission in April, carrying a central truss segment and mobile transporter. One of the crew, Jerry Ross, became the first person to make seven spaceflights—all on the Shuttle.

Also in April there was another Soyuz TM swap-over, followed by a Shuttle mission in June carrying the fifth expedition crew; commander Valeri Korzun, Peggy Whitson, and Sergei Treschev. The mission also delivered the Mobile Base System for the Canadarm 2 system. While aboard the ISS, Whitson was named the first ISS Science Officer.

Above: NASA's James Voss consulting an atlas during his stay aboard the ISS as a member of the second expedition crew in 2001.

Left: Space Shuttle astronauts Tom Jones and Ken Cockrell enter the U.S. *Destiny* Laboratory Module during an assembly mission in 2001.

Opposite: A Space Shuttle astronaut on a robot arm works outside the ISS during early assembly work of the orbiting space base.

Dragon in Space

A third nation will enter the manned spaceflight league in 2003. The People's Republic of China will launch two "taikonauts" into orbit aboard the fifth Shen Zhou spacecraft, using a Long March 2F booster from Xichang. Shen Zhou is variously translated as Divine Craft, Vessel of the Gods, or Divine Mechanism.

In April 1970, China became the fifth nation, after the Soviet Union, United States, France, and Japan, to fly its own satellite, when Dong Fang Hong 1 was launched by a Long March 1 booster. Since then, China has launched a series of scientific and national communications satellites.

The country even entered the commercial satellite launcher market, flying communications satellites into geostationary transfer orbit, mostly for American customers. Failures, including a Long March 2E that veered into a village near the Xichang launch site killing almost a hundred people, reduced China's commercial launch credibility.

Since 1976, remote sensing and materials science spacecraft have also been launched and successfully recovered, giving China experience in the re-entry and landing technologies required for manned spaceflight. The manned spaceflight plan was called Project 921 and established in 1992. Shen Zhou is very similar but larger than the Russian Soyuz spacecraft, with a forward orbital module, descent capsule, and instrument section.

Innovative design

The Chinese links with Russia included the training of two taikonauts at the Star City center, who will be involved in the project either as crew or training managers. The main differences from Soyuz are that the orbital module is cylindrical and has its own pair of solar panels, and can be fitted with a docking system. Experiment pallets can be mounted onto the outside.

The orbital module can also be left in orbit to fly independently as a mini-space station. Indeed, by docking two Shen Zhou spacecraft together, with a full complement of three crew in each, a national space station could be operated, with the two modules being left docked together after the crews have returned to Earth.

Shen Zhou weighs 16,740 lb (7,600 kg) with a

Right: The Chinese Long March 2F booster on the launch pad with its Shen Zhou spacecraft and launch escape system payload.

total length of 28 ft (8 m), and solar panel span of 64 ft (19 m). The rear instrument module with engine weighs 6,600 lb (3,000 kg), the 6.5 ft (2 m) long, 8 ft (2.5 m) diameter re-entry capsule 6,830 lb (3,100 kg), and the 7 ft (3.2 m) by 4.8 ft (2.2 m) orbital module, 3,300 lb (1,500 kg).

The first Shen Zhou spacecraft was launched into orbit by the upgraded Long March 2E booster, redesignated the 2F and equipped with a launch escape system, on November 20, 1999. The craft was placed into a 121 miles (196 km) by 200 miles (324 km), 42° inclination orbit. Supporting the mission were four Yuan Wang tracking ships located across the globe.

The spacecraft's descent capsule was safely recovered after a flight of 21 hours and 11 minutes, and 14 orbits. The orbital module remained in space. The capsule's landing was not perfect and could have possibly injured a crew member.

The next flight was delayed until January 9, 2001, and carried a monkey, dog, and rabbit to test the life support systems. Shen Zhou 2 also performed three orbital maneuvers and landed seven days after launch. It is understood, but not confirmed officially, that the capsule was badly damaged. The orbital module remained operational in space.

This mission was the most ambitious space science flight conducted by China, and included over 60 experiments in life sciences, materials processing, and crystal growth. The craft also carried cosmic ray and particle detectors. The experiments were carried in the descent capsule and orbital module, and mounted on pallets on the orbital module and instrument section.

The third Shen Zhou flight was launched on March 25, 2002, and was highly successful, leading project officials to plan one more unmanned flight, possibly in late 2002, and the first manned flight in 2003, on Shen Zhou 5.

Above: The launch of an unmanned Shen Zhou test flight from Xichang.

Left: The Shen Zhou re-entry capsule after its parachute landing in the Gobi Desert. Notice the similarity to the Russian Soyuz.

A Perilous Occupation

The first 40 years of manned spaceflight have been astoundingly successful, but space travel is by its very nature highly dangerous. It is all too easy to overlook the fact that astronauts and cosmonauts subject themselves to great risks every time they fly. Over the decades, several have been killed and many more have come extremely close to losing their lives.

Above: Russia's Sergei Avdeyev, the most experienced space traveler in the world, with 748 days in total.

Opposite: A close-up of the Russian Mir space station shows damage to the *Spektr* module after the collision of a Progress tanker.

Right: Gemini 8's commander Neil Armstrong and pilot David Scott almost lost their lives in March 1966.

At the time of writing in September 2002, there had been 233 manned spaceflights since April 1961, 140 by the U.S. and 93 by the Soviet Union or Russia. This number does not include 13 American X-15 rocket plane astro-flights which exceeded 50 miles altitude.

The number of people who have flown in space is 419, with two men having flown seven missions and six women five each. Citizens from 30 countries have flown into space, 263 from the U.S., 95 from the former Soviet Union, 11 from Germany (including one former East German), nine from France, eight from Canada, five from Japan, four from Italy, and two from Bulgaria. Countries with one space traveler are Afghanistan, Austria, Belgium, Czechoslovakia, Cuba, Hungary, India, Mexico, Mongolia, Netherlands, Poland, Romania, Saudi Arabia, Slovakia, Spain, Switzerland, Syria, the UK, and Vietnam.

The longest spaceflight—437 days—was made by the Soviet Union's Valeri Poliakov, while America's Carl Walz and Dan Bursch have flown for 195 days. NASA astronaut Shannon Lucid has flown a 188-day flight. Russia's Sergei Avdeyev has clocked up 748 days' space experience on three missions, Walz, 230 days on four, and Lucid 223 on five.

Given those statistics (and excepting February 2003's *Columbia* disaster), it is remarkable that only 11 people have been killed in flight; seven of them during the *Challenger* accident in 1986. An X-15 pilot was also lost during the last astro-flight in 1967, while three astronauts have been killed on a ground test—the Apollo 1 crew, in January 1967. The four who lost their lives in flight were Soviet Soyuz 1 pilot, Vladimir Komarov, who was killed when his craft's parachute failed after returning from orbit in April 1967, while the Soyuz 11 crew, Georgi Dobrovolsky, Viktor Patsayev, and Vladislav Volkov, were killed in space when their spacecraft depressurized while returning to Earth in June 1971.

Frequent emergencies

Several astronauts and cosmonauts have been killed in ground and air accidents, and there have been many close shaves during actual spaceflights, with several Soviet incidents only recently coming to light. The first of these was on the first ever orbital flight. Yuri Gagarin's Vostok capsule initially failed to separate from its service module during re-entry in April 1961, and could have burned up. Gus Grissom almost drowned at the end of his sub-orbital Mercury flight in July 1961, while the first spacewalker, Alexei Leonov, almost failed to get back inside his Voskhod after the first spacewalk in March 1965.

Neil Armstrong and David Scott nearly blacked out during the Gemini 8 spin in March 1966, and the Apollo 13 crew, James Lovell, Jack Swigert, and Fred Haise, just made it home after the April 1970 saga. In April 1975, following a launch abort, the Soyuz 18-1 capsule slid down a mountain side, only coming to a stop when the chute snagged a tree. Vasili Lazarev and Oleg Makarov barely escaped with their lives.

Tom Stafford, Deke Slayton, and Vance Brand were nearly gassed to death by a nitrogen tetroxide leak on landing following the Apollo-Soyuz link-up in July 1975. The Soyuz 23 crew, Vyacheslav Zudov and Valeri Rozhdestvesky, were nearly lost in a frozen lake after an aborted flight to

Salyut 5 in October 1976.
The Soyuz 33 crew, Nikolai
Ruckavishnikov and Georgi Ivanov,
aborted a Salyut 6 approach due to an
engine failure, and the mission became the
Soviet Union's "Apollo 13."

There have been several launch pad aborts
in the Space Shuttle program, in which the main
engines are shut down just before lift-off when a
fault is detected. One of these, STS-68 in August
1984, came very close to launch at T-1.9 seconds.
Had lift-off taken place and the engine then shut
down, there would have been a "contingency
abort," which has no guarantee of success.

Perhaps the most serious non-fatal
events were aboard the Mir space
station when an oxygen candle
exploded in February 1997, and
in June 1997 when a Progress
M tanker collided with
Mir's *Spektr* module.

CHAPTER EIGHT

INNER PLANETS

Mercury, another "moon"

The 3,030 miles (4,880 km) diameter Mercury is the solar system's innermost planet. The first attempts to map Mercury using a telescope were made in 1881, but were not successful. In fact, nobody knew what Mercury looked like until a spacecraft, Mariner 10, flew past the planet in 1974. Scientists were in for a surprise.

Below: Mariner 10 found Mercury to be much like the moon in appearance.

Mercury is in an elliptical orbit which takes it to within 28.5 million miles (45.9 million km) of the sun and as far away as 43.29 million miles (69.7 million km). In the non-atmospheric black sky of Mercury the sun is twice the size as it is in the Earth's blue sky. The planet rotates every 58 days and it orbits the sun once in 87 days. This means that any given place on the planet is in daylight or darkness for 176 days at a time.

Very little was known about Mercury until the planet came up with some surprises for NASA's Mariner 10, which was launched on November 3, 1973. The spacecraft was first sent to Venus and used this planet's gravity as a sling-shot to divert it onto a path toward Mercury. Mariner 10 made its first encounter on March 29, 1974, coming to within 436 miles (703 km) of the planet.

Flying in the approximate orbit of Mercury, Mariner was able to fly past the planet two more times, on September 21, 1974 and March 16, 1975, when it came its closest, at 203 miles (327 km). Mariner 10 weighed 1,108 lb (503 kg) and had an octagonal-shaped bus 11.8 ft (4.6 m) high and 4.52 ft (1.38 m) in diameter, with two solar panels, generating 820 watts of electricity. The craft was dominated by two TV cameras, looking like a pair of eyes, which generated 700-line images that were transmitted to Earth.

Mercury astonished scientists, who thought they were looking at the moon! Craters up to 125 miles (200 km) wide, ridges, lava-flooded areas, and a mountain-like ring of an impact basin, 800 miles (1,300 km)

in diameter and later called Caloris, came under view. Other measurements confirmed previously estimated temperatures ranging from –180°C (–292°F) to 430°C (806°F)—hot enough to melt lead—and a metallic core that comprises 80% of the planet.

Long-awaited return

Mercury has not been explored since Mariner 10's mission. The next spacecraft to visit it will be the Mercury Surface Space Environment Geochemistry and Ranging craft, *Messenger*, which will reach the planet in January 2008, 34 years after Mariner 10. *Messenger* will first fly past the planet and then begin to orbit it in September 2009 at a distance of 125 miles (200 km) by 9,430 miles (15,190 km), at an inclination of 80°. Getting to Mercury will involve a series of gravity-assisted fly-bys of Earth, two of Venus, and then two of Mercury itself,

before reaching it later at the right approach direction and speed. *Messenger* will carry a suite of instruments, including a camera, spectrometers, and a magnetometer, and it is hoped that it will operate for at least a year. The regular observations by *Messenger* will provide opportunities for global mapping and detailed analysis of the surface, interior, atmosphere, and magnetosphere.

The European Space Agency is also planning a Mercury explorer in conjunction with Japan, called *Bepi Colombo*, which will be launched in 2009. Using a solar-electric ion propulsion system, and protected against 400°C temperatures, *Bepi Colombo* will orbit Mercury in 2011 and will deploy a sub-satellite to study the magnetosphere and also a small lander. The lander will carry a surface penetrator which will send back data on the structure of the planet.

Above: Mariner 10 made three passes of Mercury in 1973–74, the only craft to explore the innermost planet.

Venus—Paradise Lost

Three Soviet and one U.S. spacecraft were aimed at Venus, but failed, before NASA's Mariner 2 success. Mariner 2 was not just the first spacecraft to explore Venus, it was the first to explore *any* planet, reaching Venus at the very early date of December 14, 1962.

The 447 lb (203 kg) spacecraft, which was launched in August 1962, was based around a pyramid truss frame 9.9 ft (3 m) high, with twin solar panels 4.9 ft (1.5 m) long, providing up to 222 watts of electricity.

It was already known that the planet was shrouded in thick cloud, and many scientists wondered whether underneath lay a lush tropical paradise. It was this possibility that motivated both the Americans and Russians to explore the planet so early in the Space Age.

Mariner 2's radiometers caused the first upset when they indicated that the surface temperature is 460°C (860°F). The carbon dioxide cloud cover—which Mariner found to be at its thickest between 50 miles (80 km) and 34 miles (56 km) altitude—was causing an extreme greenhouse effect. The veil of the mysterious Venus had been lifted: it was a hellish world rather than a paradise. Despite this, not all scientists believed Mariner's readings, and some remained optimistic.

The thick cloud covering of Venus makes it extremely reflective to the sun's light, so the planet is the brightest object in our sky, after the sun and the moon. The diameter of Venus is 7,518 miles (12,098 km). The planet makes an orbit of the sun in 224 days but rotates only once every 243 days. Its maximum distance from the sun is 67 million miles (109 million km), and the closest is 66 million miles (107 million km).

The Soviet Union was the early star performer on Venus missions, despite several failures. It was aiming for a landing on the Venusian surface and was very persistent about it. The Russian spacecraft Zond 1 failed in 1964. Zond 3 was flown past the moon to test a Venus imaging system in 1965, while Veneras 2 and 3 were unsuccessful in 1969.

Melted and crushed

The first spacecraft to penetrate the Venusian atmosphere was Venera 4, launched by the Soviet Union on June 12, 1967. NASA's Mariner 5 was launched two days later on a fly-by mission. The 2,436 lb (1,106 kg) instrumented Venera 4 craft carried a 3 ft (1 m) diameter, 843 lb (383 kg) landing capsule which was released as the main craft plunged into the atmosphere. The capsule, designed to withstand 350G deceleration, reached speeds of 6.2 miles per second (10 km/ps), and its ablative

Right: Mariner 2 was the first planetary explorer, passing Venus on December 14, 1962.

followed in 1969. Their 892 lb (405 kg) capsules had been strengthened to withstand up to 27 Earth atmosphere pressure, and Soviet scientists were confident of success. On May 16, Venera 5's capsule, descending under its single parachute, transmitted data for almost 50 minutes when at 61 miles (26 km) altitude, the craft's internal temperature rose to 280°C (540°F) and the pressure rose to 27 atmospheres. Transmissions ceased at 53 minutes.

Venera 6's capsule began its suicide plunge on May 17, and after 51 minutes' transmission was indicating 26 atmospheres. Transmissions ceased at 7 miles (12 km) altitude. Extrapolations of data from the two Veneras indicated that surface temperatures and pressures would be 400°C (750°F) and 60 atmospheres.

What would Venera 7 find?

Left: Mariner 2 is launched aboard an Atlas Agena booster from Cape Canaveral.

Below: The Soviet Union's second series of Venus explorers aimed to land a capsule on the surface, but none were fully successful.

heatshield faced temperatures of up to 11,000°C (20,000°F).

Despite these tolerances, the craft did not survive to hit the surface under its single parachute, which was itself designed to withstand 450°C (850°F) temperatures. Transmissions ceased at an altitude of 16.8 miles (27 km), where the craft had measured the atmospheric pressure as 22 times that of the Earth's and the temperature at 280°C (540°F). The main discovery made by Venera 4 is that the atmosphere consists of up to 95% carbon dioxide.

The similar Veneras 5 and 6

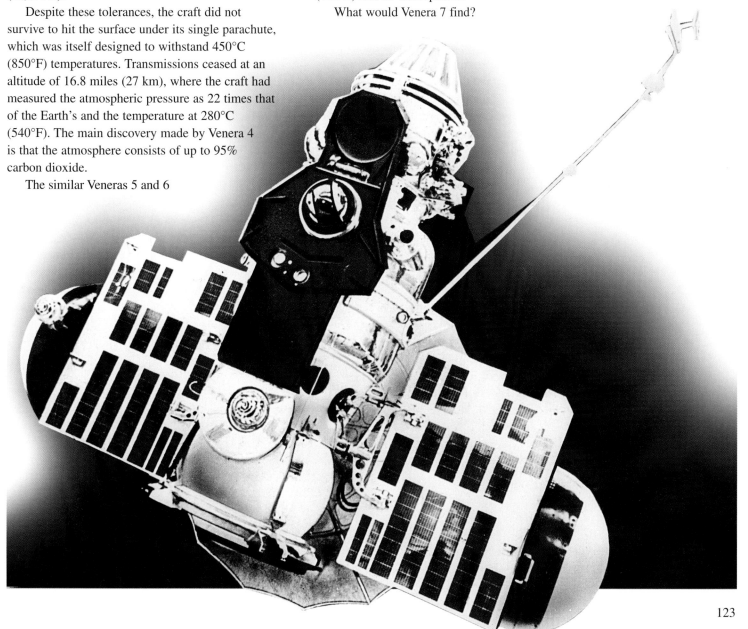

The Venusian World

On December 15, 1970, the first certain contact with the Venusian surface was made by the Soviet Venera 7 capsule. After so many attempts, the 1,100 lb (500 kg) capsule just had to succeed, so scientists built it to withstand temperatures of 540°C (1,000°F) and 180 atmospheres!

Succeed it did. The capsule was retained by the mother ship much longer than in previous missions, before being released, so that it was attached to the main cooling system for a longer period. The single parachute deployed at 37 miles (60 km) at a speed of 447 mph (720 km/h), and all seemed well. Thirty-five minutes later the transmissions ceased. All seemed to have been lost.

However, after analyzing the received data stream, scientists discovered 23 minutes of 1% normal strength signals, proving that Venera 7 had reached the surface intact at 5°S, 351° longitude. Data indicated approximate pressures of 90 atmospheres and temperatures of 475°C (890°F). An improved Venera 8, equipped with a photoresistor to measure light levels at the surface, followed with a landing on July 22, 1972.

Veneras 9 and 10 were to be of a completely different design. Both were launched in June 1975, the craft reaching their destinations on October 15 and 22 respectively, marking the flights of the first Venus orbiters and the return of the first images of the Venusian surface.

Venera 9 deployed its lander before entering orbit. The 1,674 lb (760 kg) lander reached 39 miles (64 km)

Below: The surface of Venus as seen for the first time, from the Soviet probe Venera 9 in October 1975.

altitude and deployed the upper section of its aerothermal sphere, deployed a braking chute and the 9.1 ft (2.8 m) diameter main chutes.

At 30 miles (50 km), the chutes were jettisoned and the craft made free-fall descent through cloud layers, landing at 5 mph (8 km/h) on a 15° slope at the base of a hill near Beta Regio, at 32°N/291° longitude, cushioned by a donut-shaped air bag.

A hellish surface

Instruments measured 1.2 mph (2 km/h) winds, a temperature of 460°C (860°F), and pressure of 90 atmospheres. Transmissions lasted for 53 minutes, during which time the data returned included a black and white image of the surface showing a rocky area with 11-15 in (30-40 cm) flatish, sharp rocks in light which one of the scientists compared to "Moscow on a cloudy day."

Venera 10 landed on a 9.8 ft (3 m) sloping slab of rock in a landscape of flat rocks at 16°N/291° longitude, with temperatures at 465°C (870°F) and 92 atmospheres' pressure. The transmissions lasted for 65 minutes.

NASA launched two Pioneer Venus spacecraft in 1978, one of which entered orbit to map the planet using radar, the other to deploy one 696 lb (316 kg) and three small 204 lb (93 kg) probes into the atmosphere, later hitting the surface at the end of their missions on December 9. The first detailed atmospheric model of the Venusian atmosphere resulted—including the conclusion that the clouds are composed largely of sulfuric acid droplets and that the clouds clear at about 18 miles (30 km) high, but the sky is a gloomy red murk

Veneras 13 and 14, launched in 1981, returned the first color pictures of the surface and performed the first soil analysis, which indicated a high basalt content. The amount of sunlight reaching the surface was measured at between 2.4% and 3.5% and the images showed orange-brown rocks, an orange sky, and a horizon that seemed 100 miles away due to a mirage effect.

The Soviet Union's exploration of Venus ended in June 1985 when two Vega spacecraft en route for a rendezvous with Halley's Comet dropped off two Venera-like landers, which deployed Teflon-coated plastic balloons inflated with helium. At the end of a 42.6 ft (13 m) tether below was a three-section gondola with nine instruments, which returned more data.

Left: One of the twin Pioneer spacecraft launched to the hot planet in 1978, which together deployed four probes to explore the Venusian atmosphere.

Below: The launch of the Pioneer Venus orbiter aboard an Atlas Centaur booster from Cape Canaveral.

seen from ground level.

The Soviet Veneras 11 and 12 were also launched in 1978, performing repeat missions of their predecessors, but no pictures were returned and much data was lost, possibly due to electrical storms on the surface. The reverberations of a 15-minute thunderclap was recorded by Venera 12.

Venus Unveiled

Thick cloud cover had made it impossible to take a global view of Venus's surface—until the introduction of space radar imaging. Signals sent from a spacecraft onto the surface of a planet bounce back, and measurements provide data on the profile of the surface.

Above: A computer generated image of the surface of Venus made up of images taken by the Magellan radar mapper.

The first radar mapping of the planet was made by Pioneer Venus 1, which was launched in May 1978 and entered orbit around the planet in December. The 1,219 lb (553 kg) craft was based on an 8.29 ft (2.53 m) diameter, 4 ft (1.22 m) high cylindrical, spin-stabilized bus. Its 14,580 solar cells provided 312 watts of power.

The craft's solid propellant retro-rocket placed it into a 24-hour orbit around Venus with an eventual low point, or perigee, of 93 miles (150 km), and an apogee of 41,805 miles (66,889 km).

The radar mapper allowed a global, topographical map of an area between 73°N and 63°S to be made, with a resolution of 46 miles (75 km). The images revealed an extraordinary scene of a tortured surface, featuring two large continents,

called Ishtar Terra and Aphrodite Terra, the size of Australia and Africa respectively, and a 6.5 mile (10.8 km) high extinct volcano, called Maxwell Montes.

Russia's Venera 15 and 16 radar mappers, launched in 1983, showed a more chaotic terrain, which whetted appetite of the U.S. scientists who were planning to launch the first dedicated Venus radar mapper—*Magellan*. The Space Shuttle *Atlantis*/STS-30 deployed *Magellan* in Earth orbit after launch on May 4, 1989, after a postponement from 1986, due to the grounding of the Shuttle fleet after the *Challenger* accident.

The 7,594 lb (3,445 kg) spacecraft, 21.6 ft (6.46 m) high, was dispatched from Earth orbit by an upper stage. *Magellan* cruised through space in solar orbit, taking it one and half times around the

sun, heading for a rendezvous with Venus in August 1990.

A solid propellant retro rocket placed *Magellan* into an almost polar orbit, enabling it to cover the whole planet in three years as Venus rotated very slowly beneath it. The resolution of the *Magellan* radar images was between 393 ft (120 m) and 983 ft (300 m).

Shifting surface

The synthetic aperture radar produced 1,852 images in 9,934 miles (15,996 km) long latitudinal swaths between 10.55 miles (17 km) and 17.39 miles (28 km) wide. The swaths were compressed into mosaic maps of the surface. Surface feature height was measured to a 98 ft (30 m) accuracy by an altimeter. The radar and altimeter data was combined with radio measurements.

The chaotic terrain of Venus seems to have been caused by recent volcanic activity, which resurfaced the planet. The low number of impact craters appears to confirm this. *Magellan* images revealed a wide range of features: impact craters with bright ejecta material, fractured plains, volcanoes, lava flows, and highlands on Ishtar Terra.

Magellan had completed four 243-day cycles by May 1993. It was then used to perform maneuvers to demonstrate the use of the Venusian atmosphere to slow down and maneuver a spacecraft. This is called aerobraking. In October 1994, *Magellan* started to enter the Venusian atmosphere, plunged inside at high speed, and burned up.

No other spacecraft are planned for specific exploration of Venus, although the European Space Agency has considered a Venus orbiter. However, planetary probes such as *Galileo* when it was en route to Jupiter, and *Rosetta* which will be launched to explore a comet, have used or will use Venus for slingshot maneuvers on their way to other destinations in the solar system.

Left: Another computer-generated image shows a close-up of the surface.

Below: NASA's Venus *Magellan* radar mapper completed four 243-day exploration cycles before entering the atmosphere in 1994.

127

CHAPTER NINE

MARS

Martian Myth

Mars has fascinated mankind since prehistoric times. Its distinctive red color when seen in the night sky, which gave rise to its name of the Red Planet, contributes to this fascination. Early observations by telescopes revealed tantalizing features. Parts of Mars seemed to change shape and color, and lines were seen on the surface. In the 19th century, speculation that these features were canals led to suggestions that the planet may be inhabited. In the early 1960s, both American and Soviet scientists were eager to discover whether there could be any truth to these assertions.

Right: The Mariner 6 and 7 spacecraft started to unravel the mysteries of Mars with high resolution images showing the multi-featured surface.

Below: The legendary image taken by Mariner 4 in July 1965 showing that the surface of Mars is cratered.

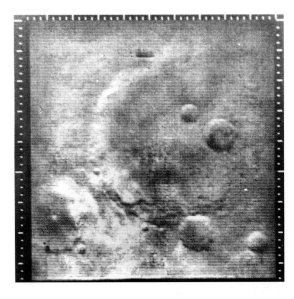

The idea that there may be life on Mars, encouraged by books by H.G. Wells and the science fictions films of the 1950s, still goes on today—although following the Viking missions of the 1970s (*see following pages*), most people accept that there can be no advanced life on Mars.

Mars orbits the sun at a maximum distance of 154 million miles (249 million km), and minimum 128 million miles (207 million km) every 687 days. It is 4,215 miles (6,787 km) in diameter, and rotates in almost the same time as the Earth does, in 24 hours and 37 minutes. The maximum surface temperature on Mars is 18°C (64°F) on the equator in summer, and it has a thin atmosphere of carbon dioxide.

The planet has two moons, Phobos and Deimos. Phobos is 16 miles (27 km) long and 12 miles (19 km) wide. Deimos is 8 miles (12 km) long and 6.2 miles (10 km) wide. Phobos is in an unusual orbit. It crosses the Martian sky in only four and a half hours, from west to east, and appears again 11 hours later. Deimos, in a deeper orbit, remains in the sky for two and a half Martian days.

Six attempts to fly a craft to Mars failed between 1960 and 1964, before the American Mariner 4 was launched in November 1964 and flew past the planet on July 15, 1965. It caused quite a stir.

The spacecraft bus was similar to that which later flew on Mariner 10 to Mercury, measuring 4.32 ft (1.38 m) in diameter and equipped with four solar panels comprising 7,000 solar cells, producing 700 watts of power. Mariner 4 carried a single TV scanning camera which was programmed to return just 21 images as it flew past the Red Planet at a distance of 6,000 miles (9,600 km), covering a swath from 37°N to 55°S.

Each image comprised 40,000 elements transmitted as numbers according to the amount of light received by the camera. Other instruments returned data on the atmosphere. The early processed images were quite fuzzy, and it seemed that the mission would be a disappointment. Then came the legendary image No 11, which showed that Mars looks quite similar to our moon. The surface could be seen in relatively fine detail, revealing the remains of a large crater, and within it several more craters, in a region later called Atlantis.

Freezing desert

Mariners 6 and 7 were launched in February and March 1969 to make a more detailed survey during a closer, 2,130 mile (3,430 km) fly-by. The craft looked similar to Mariner 4 but were larger, weighing 909 lb (413 kg), and equipped with a more powerful camera system. The images showed a chaotic planet not just with hundreds of craters, some filled with carbon dioxide frost, but also revealing a moon-like globe with desert-like flat areas and a frozen carbon dioxide south polar cap. Among other instruments was a radiometer which measured the temperature on the equator as –73°C (–100°F), and at the pole, –125°C (– 193°F). What was needed now was a more continuous survey by an orbiter.

The 1,245 lb (565 kg) Mariner 9 was launched in May 1971 (Mariner 8 failed), and became the first Mars orbiter on November 14, entering an 80° orbit, which would increase the coverage to 70% of the entire surface. Initial images indicated that Mariner 9 had arrived at the time of a huge dust storm, which obliterated the surface until January 1972. During this time the craft took some photographs of the Martian moons, Phobos and Deimos, showing them to be pockmarked by craters. Scientists concluded that they were probably asteroids which had been captured by Mars' gravity.

Had Mariner 9 been on a fly-by the mission, it would have been a waste of time, but as the dust cleared, Mariner's panorama proved to be spectacular, revealing the diverse surface indicated by the earlier Mariner 6 and 7 fly-bys. Mariner 9 photographed what appeared to be dry river beds, indicating that water once flowed on the surface, and that there are huge valleys, including the 2,500 miles (4,000 km) long, 60 miles (100 km) wide Valles Marineris. It also discovered volcanoes, including spectacular Olympus Mons. This huge volcano is much higher than the Earth's Mount Everest, towering 15 miles (22 km) into the Martian sky, and is about 340 miles (550 km) in diameter. It is made up of thousands of individual lava flows which can be traced for hundreds of kilometers across the surrounding terrain.

While Mariner 9 was orbiting, the first Soviet orbiters, Mars 2 and 3, reached the planet. In 1973, Mars 5 also entered orbit. These spacecraft returned similar but lower quality data.

Below: An historic image taken by the first Martian orbiter, Mariner 9, showing the Martian volcano Olympus Mons, five times higher than Mount Everest.

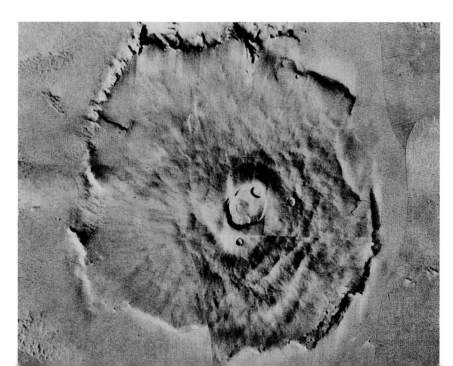

The Vikings

By the mid-1970s NASA had launched two spacecraft capable of landing on the surface of Mars. Vikings 1 and 2 reached the planet on July 20 and September 3, 1976 respectively. The identical Viking spacecraft were spectacular successes, providing thousands of color images taken from the surface, and also included two orbiters which continued the intensive survey of Mars started by Mariner 9.

T he landers were encased in a lens-shaped 806 lb (366 kg) aeroshell which plunged into the upper Martian atmosphere, protected against 1,500°C (2,700°F) temperatures. A 53 ft (16.2 m) diameter parachute opened at about 3.1 miles (5 km) altitude, the aeroshell was jettisoned, and the landers' legs deployed.

At about 4,590 ft (1,400 m) above the surface, the crafts' three throttleable engines and four thrusters ignited, heading for a touchdown at about 7.8 fps (2.4 m/s).

The craft were basically six-sided aluminum-titanium buses housing external instruments, including a meteorology boom, with a maximum height of 6.8 ft (2.1 m). Each of the three landing legs had footpads, and the craft were powered by two radioactive plutonium oxide radioisotope thermoelectric generators.

The landers were equipped with a 9.8 ft (3 m) robot arm with a scoop for collecting soil, which was deposited into an internal laboratory comprising a biology distributor, gas chromatograph, mass spectrometer, and an X-ray spectrometer. Ingenious methods were devised to detect possible signs of life on Mars, including heating the samples, then adding water and nutrients to culture any living organisms such as bacteria. However, no conclusive proof of life was found by either of the landers.

The craft also returned color images of the rust-colored surface, showing rocks up to the horizon, a pink sky, sunsets forming concentric arcs in the sky, and carbon dioxide frost on the ground. Over 4,500 images were transmitted from the landers. The carbon dioxide atmosphere was found to reach a surface pressure of 7.6 mb, dropping by 30% in winter. The temperature was about –33°C (–27°F) in mid-afternoon, and wind speeds of up to 40mph (51 km/h) were detected.

Viking 1 landed in a barren desert landscape strewn with rocks that looked like remnants of a lava flow, while Viking 2 touched down on a flatter,

Below: The robotic arm of the Viking spacecraft, which collected samples of Martian soil to be analysed aboard the spacecraft.

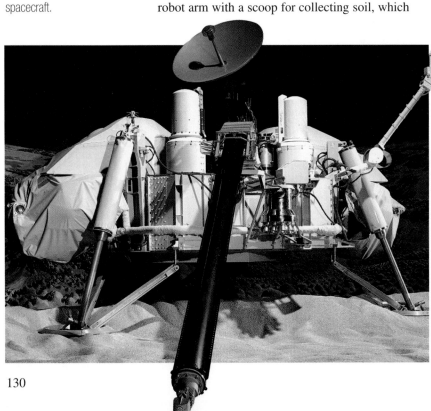

rockier site, possibly featuring rocks ejected from the Mie crater 60 miles (100 km) away. The landings by Viking 1 and 2, at Chryse Planitia and Utopia Planitia respectively, are considered to be a major milestone in space exploration, but for various reasons—mainly budget restrictions and the apparent lack of Martian life—were not followed until 1997, when Mars Pathfinder captured the imagination of the world in another age.

Mars almost entirely mapped

The Viking orbiters were as successful as the landers. By the time they stopped transmitting, they had taken thousands of images of the Martian surface. Their most important achievement was to enable mission managers to take a close look at the planned Viking 1 and 2 landing sites and, as a result, change the targets to what looked like smoother areas, thus probably saving the two landing missions. Viking 1 was to have landed about 3° south of where it eventually put down in Chryse Planita, while Viking 2 was to have landed in Cydonia. Both areas are strewn with boulders and obstacles. However, as a result of the change of the Viking 1 landing site, the U.S. was robbed of a July 4 bicentennial landfall in a new world.

The orbiters returned over 51,000 images, mapping 97% of the surface of Mars at 98 ft (30 m) resolution, and 2% of the surface at 81 ft (25 m), or better. Images of features above 30° latitude showed soft outlines and profiles, possibly indicating sub-surface water ice—a discovery that was to come to the fore in a big way in the next century.

Above: Chryse Planita by day as seen by the camera aboard the first Mars lander, Viking 1, in 1976.

Left: Viking 2's view of Utopia Planita shows a desert-like orange surface with rocks and boulders.

A Sojourn at Ares Valles

In the early 1990s, NASA was planning a network of Mars surface environmental survey craft. A single technology demonstration mission was planned. However, the network was canceled and the single craft was renamed Mars Pathfinder. This became part of NASA's low-cost Discovery program, introduced as a result of budget restrictions.

Another technology program, the Microrover Flight Experiment, was added to fly piggyback on the Pathfinder, and was renamed Sojourner. Costing just $265 million, a fraction of what Viking had cost, the two craft were pioneers of low-cost space exploration. Due to the power of the Internet and the openness of NASA, the mission could be monitored and its images downloaded live by millions of people all over the world.

The 1,973 lb (895 kg) Mars Pathfinder and its 23 lb (10.6 kg) Sojourner payload was launched on December 4, 1996, and made a direct plunge into the atmosphere of Mars, protected by an aeroshell heatshield. After descending under a 36 ft (11 m) diameter parachute, the payload was dropped onto the surface on July 4th, 1997, protected by giant airbags.

Ares Valles, in the northern hemisphere, had been selected as the landing site because it was thought to be an ancient flood plain and one of the rockiest parts of Mars, with a wide variety of rocks deposited during a catastrophic flood. The site, at 19.33°N, 33.55°W, was close to Viking 1's landing place in Chryse Planita. Photographs taken from orbit show a spectacular view of what looks like a dry river basin with many rivulets, suggesting that the planet was at one time warm and wet, and with a thicker atmosphere.

The deflated helium bags exposed the lander to the Martian environment, and soon the microwave oven-sized, six-wheeled, solar powered, computer-controlled Sojourner rolled down the ramps from the Pathfinder and began its exploration in what was nicknamed the Rock Garden. The rover began to make short journeys, equipped with an alpha proton X-ray spectrometer and three cameras. Filmed by a camera on Pathfinder, its longest trek was a 165 ft (50 m) clockwise journey around the lander.

Exploring the neighborhood

The Sojourner camera provided super-pan images, each containing 15 frames, comprising six frames for three-color stereo and nine more individual narrow-color bands. Red, green, and blue filter images were combined to simulate slightly amplified natural color. Many features became familiar in the daily scenes, such as the rock, named Yogi, Mermaid Dune, and the rim of an impact crater, called Big Crater, visible on the horizon.

The Pathfinder, renamed the Sagan Memorial Station after Carl Sagan, was fitted with an instrument to measure the magnetic properties of the soil, as well as windsocks, and a package to measure the atmosphere and make meteorological readings.

From landing to shutdown on September 27, Pathfinder returned 2.3 billion bits of data, including 16,500 images, while the Sojourner took 550 images and made 15 chemical analyses of rocks, and measured winds.

Some of the samples found rock chemistry different than that in the so-called "Martian meteorites" discovered on Earth, which grabbed world media attention when some scientists alleged they came from Mars and contained micro-fossils of ancient Martian life. It is now considered doubtful whether these meteorites did originate from Mars, let alone whether they contain genuine fossils of ancient life. Sojourner found basaltic andesite content, similar to that seen at the Viking 1 and 2 sites.

Temperatures were found to be about 10°C (18°F) warmer than those measured by Viking 1, while frequent dust devils occurred, which in turn caused wind-abrasion on rocks and also produced small dunes.

Pathfinder exceeded its predicted design life by a factor of three and the Sojourner, dying on October 6, by 12 times. By the time that Pathfinder stopped transmitting it had returned 83% of a 360° high-resolution super-pan, of the terrain, showing hills in the distance and the tiny rover dwarfed by large rocks.

Opposite: A Delta II booster launches Mars Pathfinder in December 1996, en route to its spectacular mission at Ares Valles.

Left: The first and so far only Mars rover, the Sojourner, was the size of a microwave oven. It took over 500 close-up images.

Below: A panoramic super-pan view of the Martian surface taken by Mars Pathfinder after its Independence Day landing in 1997.

The Search for Water

If there is—or was—life on Mars, water is a vital ingredient. However, if there is water, it doesn't necessarily mean that there is life. The missions of NASA's Mars Global Surveyor and Mars Odyssey certainly went a long way to proving that there used to be a great deal of water on the surface, and that much of it may still exist underground. As for life, we will have to wait for some samples to come back to the Earth.

After the loss of Mars Observer in 1993, NASA introduced the Mars Surveyor series of missions, consisting of orbiters and landers, starting with Mars Global Surveyor (MGS), which would carry five of the seven experiments flown on the Observer.

MGS kicked off the Mars Surveyor program with a launch on November 7, 1996, but soon after reaching space, one of the craft's two solar panels failed to fully deploy, and there were fears that the mission would fail as a consequence. However, troubleshooting overcame the partial loss of power, and MGS arrived in an initial Mars orbit with the help of a retro burn on September 11, 1997.

The spacecraft subsequently used the Mars atmosphere as a brake to gradually reduce the orbit, but over a longer period than planned for fear of damaging the errant solar panel, in a series of aerobraking maneuvers, which save fuel.

MGS proved to be a great success, returning high-resolution images of a variety of features on the whole of the Martian surface, thanks to its orbital coverage. Many of the images clearly show evidence that there was running water eons ago, and also remarkable views of features such as the polar caps and the familiar sites, such as the Valles Marineris and Olympus Mons.

Although retired in 2001, the spacecraft continued to operate, returning images to complement the new Mars Odyssey, which arrived in orbit on October 24, 2001. Meanwhile, a Japanese spacecraft, Nozomi-Planet B, was launched in July 1998 to orbit Mars in 1999, but due to a lack of velocity the spacecraft will not arrive until November 2003.

Mars Odyssey was to map the amount and

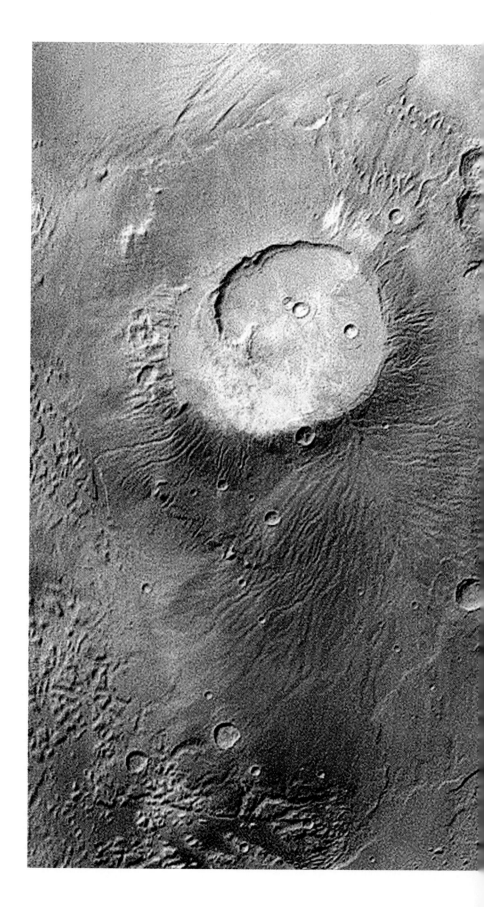

The Mars Global Surveyor conducted a detailed survey of almost the whole Martian surface from orbit. These images show the planet's immense variety of landscapes, including (opposite) the crater Apollinaris Petra with an icy fog cloud, (left) gullies showing evidence of once-running water. and (below) exquisite shapes in the polar cap.

climate and geology, and to "prepare for human exploration." In May 2002, NASA announced that *Odyssey* had found "water ice in abundance under Mars' surface." The popular media then went hysterical about the possibility of life on Mars. NASA said that the gamma ray spectrometer "indicated the presence of water ice" in the upper meter (3 ft) of soil in a large region surrounding the planet's south pole.

The scientific team came to a "conclusion" over the evidence. This indicated that there was enough water ice to fill Lake Michigan twice over. "It may be better to characterize this layer as dirty ice rather than ice containing dirt," said one scientist. Nevertheless, the finding was not a surprise since the evidence on the surface was very clear.

The positive result of this finding for future exploration is that a human expedition could have a plentiful supply of processed water and processed hydrogen as a fuel for the spacecraft's return journey to Earth. *Odyssey* will continue to survey Mars to look for other potential reservoirs of water. Meanwhile, other spacecraft will also continue the search for water, the Martian holy grail.

distribution of chemical elements and minerals that make up the Martian surface and sub-surface, looking especially for hydrogen. This would most likely be in the form of water ice in the shallow subsurface, and would be detected using a thermal emission imaging system and gamma ray spectrometer.

Preparing for a manned mission

The craft also carries a Mars radiation environment experiment and will provide communications support for some future Mars missions. As a tribute to Arthur C. Clarke, the renowned science fiction author and space visionary, the spacecraft was renamed *2001 Mars Odyssey*. It got to work "following water," as NASA put it, to determine whether life arose on Mars, to characterize its

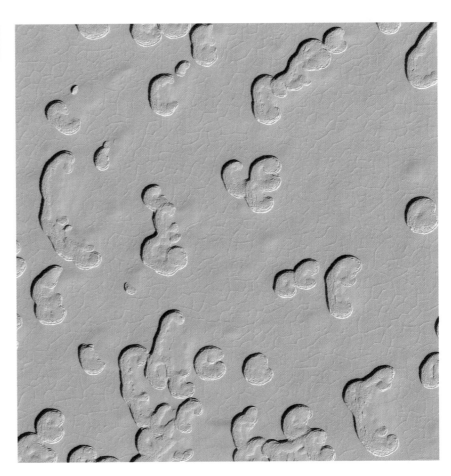

Martian Casualties

The assault on Mars—the God of War—resulted in many casualties. No planetary exploration program has suffered as many failures, and the record of the Soviet Union was quite tragic.

The Soviets made their first attempt on October 10, 1960, just three years into the Space Age, while premier Nikita Khruschev happened to be visiting the United Nations. The launch failed, as did another four days later.

A Soviet Mars probe made Earth orbit in October 1962 but failed to leave, while Mars 1, launched on November 1, successfully headed for a rendezvous with the Red Planet only for communications with it to be lost in March 1963. The craft passed Mars silently by at a distance of 120,000 miles (193,000 km) in June that year. Another Mars probe failed to leave Earth orbit in November 1962. This was followed by America's first attempt, Mariner 3, on November 5, 1964, but the craft was lost when the payload shroud failed to separate.

Contact was lost with the Soviet Zond 2 in May 1965, and in 1969, the nation suffered two more launch failures. The U.S.'s Mariner 8 also failed to launch, in 1971, a year in which an attempted Soviet Mars orbiter again failed to leave transfer orbit.

In November-December 1971, Soviet spacecraft Mars 2 and 3 successfully entered Mars orbit but were thwarted by dust storms and by the failure of both landers. Mars 2 crashed, and it is possible that the Mars 3 lander made a safe touchdown but was unable to send data owing to the sister craft's communications failure.

The Soviets launched four Mars probes in 1973, arriving in February-March 1974, only one of which, Mars 5, was successful. The Mars 4 engine failed and the spacecraft was not able to orbit Mars. Mars 6 and 7 were fly-by missions in which landers were jettisoned. The Mars 6 capsule apparently landed but may have been covered by its parachute, while the Mars 7 lander was deployed but suffered an engine failure and missed Mars altogether.

These doomed Soviet missions were followed in 1988 by the dual Phobos 1 and 2 flights to explore the Martian moon of the same name. Phobos 1 was accidentally commanded to shut down and was lost 12 million miles (19 million km) from Earth, while

Right: Two Soviet Union Phobos craft were launched to land on the Martian moon, but both failed.

contact was also lost with Phobos 2 just as it was arriving at Mars in March 1989.

The next two Mars missions to be launched were also failures. NASA's Mars Observer, launched in 1992, was lost just as it was about to enter Mars orbit in August 1993, apparently when the engine for the orbit insertion burn exploded. The international Mars 96 mission led by Russia was lost in very low Earth orbit after a botched Proton launch, re-entered the atmosphere, and crashed into the Pacific Ocean and onto parts of South America.

Mission scuppered

After the NASA Mars Pathfinder success in 1997, the space agency suffered the ignominy of two back-to-back failures, one of which was quite farcical. The Mars Climate Orbiter was launched in December 1998 and arrived at Mars on September 23, 1999, fired its engine, and flew straight into the surface, like a kamikaze bomber. Its angle of orbit had been miscalculated, due to a mix-up between English imperial and metric units by two separate

teams involved in the development of the spacecraft.

If this was not enough, on December 3, 1999, the Mars Polar Lander, the most sophisticated Mars craft since the Vikings, also arrived at Mars and suffered a suspected engine failure as it was descending to the surface. The mission also included two surface penetrator probes, but contact with these was lost.

Budget cuts in the wake of NASA's failures, and more realistic assessments of the technologies required for Mars sample return missions, resulted in the cancelation of several Mars missions. Two victims were a Mars lander-rover mission earmarked for 2003, and an orbiter-lander-rover mission for 2005 in co-operation with France, with the potential of also being a sample return mission.

The Quest for Life

The ultimate aim of unmanned Mars exploration is to return samples of the Red Planet's soil to Earth. We can speculate that there may be bacterial organisms living in the Martian soil, but it is probably only through examining samples here on Earth that we will know for sure.

This question will not be answered by just one sample return mission, but by several. NASA has conceded that the first sample return mission in unlikely to be launched before 2012, due to the usual technical and budgetary hurdles. The biggest question is whether NASA or other space agencies will be able to demand the huge budgets required, and whether these missions will be seen as priorities as the 21st century progresses.

Before any such missions are flown, several precursor flights are required to gain the technological experience of hands-on Mars exploration, to establish a communications infrastructure to support the sample return missions, and to locate potential new landing sites.

The first steps will be made in 2003 with the launch of a new fleet of orbiters and landers. The European Space Agency will launch its Mars Express orbiter, which will provide high-resolution imaging and mineralogical mapping. It will be supported by the Japanese Nozomi orbiter, which was actually launched in 1998, but due to a malfunction will not arrive at Mars until 2003 (*see page 134*).

Below: The UK's Beagle 2 spacecraft is due to land on Mars in December 2003 to analyse the soil for signs of life.

Attached to Mars Express will be a piggyback payload, Britain's Beagle 2, which will land on the surface and perform in-situ analysis using gas chromatography and spectroscopy, in the hope of finding indications of biological activity. A tiny instrument like a mole will burrow into the surface, under large boulders, and take samples from inside rocks. The samples will be analyzed for signs of organic material. The craft will look for indications of methane in the atmosphere, which could be taken—or mistaken—for signs of the existence of primitive microbes.

Comprehensive mapping

Also in 2003, NASA intends to launch two powerful Mars Rovers, which will analyze soils using five instruments while roaming 330 ft (100 m) a day, for missions lasting 90 days, during 2004. Budget cuts may reduce the number to one rover. In 2005, NASA will launch the Mars Reconnaissance Orbiter, a powerful mapping craft with the capability of a spy satellite, with a resolution of 8 inches (20 cm), in an effort to follow-up on tantalizing hints of water detected by previous orbiters.

Smart landers will be developed to demonstrate accurate landing and hazard avoidance capability, and the first of these craft could carry an advanced long-range rover which will act as a mobile laboratory. This could be launched on the first Smart Lander in 2007.

As the build-up to sample return missions continues, NASA has proposed a new line of Smart missions such as a Mars airplane, which would fly like a glider to parts of Mars that other craft cannot reach. A simple unpowered glider was designed by NASA to cruise 1,100 miles along the five-mile-deep Valles Marineris "Grand Canyon" in three hours, to deduce its history.

The flights of all these precursor missions will depend on budgets and whether the search for life on Mars is considered worth all the effort, since any life that exists can surely only be of the most elementary form.

CHAPTER TEN

GIANTS

Exploring Jupiter's Realm

The Italian Galileo Galilei made the first detailed observations of Jupiter in 1609 when he pointed his telescope at the planet. Almost four centuries later a probe named after him was sent to Jupiter, the latest in a series of spacecraft to explore the gas giant.

Right: The first spacecraft to explore Jupiter was *Pioneer 10* in December 1973, which showed the turbulent clouds of the giant planet in close-up.

The largest planet in the solar system, Jupiter orbits the sun between 460 million miles (740 million km) and 506 million miles (815 million km), and has a diameter at the equator of 88,000 miles (142,000 km). The planet is slightly squashed, so its polar diameter is 82,955 miles (133,500 km). It takes 11.86 years to go once around the sun.

Early telescopic observations located 12 moons, but many more have been found by spacecraft and by powerful modern telescopes. A swirling mass of multicolored bands of hydrogen, ammonia, hydrogen sulfide, and phosphorus gases swirl around the planet, which rotates in ten hours, the fastest rotation of any planet. Jupiter also has a spectacular storm called the Great Red Spot, which has existed for hundreds of years.

The first Jovian explorer was *Pioneer 10*, launched on March 3, 1972, and passing the planet at a distance of 80,750 miles (130,000 km) on

December 5, 1973. The 568 lb (258 kg) hexagonal-shaped craft was topped by a large antenna 8.9 ft (2.74 m) wide, but did not carry solar panels since the sun's power at the distance of Jupiter is just 4% of that which the Earth receives. Two SNAP radioisotope thermoelectric generators using plutonium 238 as fuel were mounted at the end of two booms, providing 140 watts.

Pioneer 10 transmitted over 300 high-resolution images of Jupiter. The craft suffered some damage from a large radiation belt it discovered surrounding the planet. Its mission over, *Pioneer* traveled onward out of the solar system and toward the stars. The spacecraft carries a plaque on its side which depicts two human beings and mankind's position in the solar system—just in case it is intercepted by another civilization thousands or even millions of years from now as it travels through interstellar space.

Multi-planet missions

Three other spacecraft were sent to Jupiter. *Pioneer 11*, launched on April 6, 1973, reached the planet on December 3, 1974, and was the first spacecraft to use a planet as a slingshot, flying by the south of Jupiter to be whipped northward, in the direction of Saturn.

Following *Pioneer 11* were *Voyagers 2* and *1* (*Voyager 1* was launched second) on March 5 and July 9, 1979. The 1,818 lb (825 kg) *Voyagers* were dominated by a 12 ft (3.66 m) diameter antenna for X and S-band communications, under which was mounted the spacecraft bus, extended from which were three booms and an antenna. One of the booms held most of the instruments, including wide and narrow angle TV cameras. The picture quality was much improved on earlier missions, and the most spectacular images were of Jupiter's moons, Io, Europa, Ganymede, and Callisto.

The next spacecraft to visit Jupiter was *Galileo*. The probe became the first to orbit the planet and the first to deploy a descent capsule to plunge into its thick cloud-deck. The spacecraft was deployed from the Space Shuttle *Atlantis* on October 18, 1989, and en route to Jupiter it flew past the Earth twice and Venus once to create a gravity-assisted flight path. It also became the first spacecraft to explore two asteroids in the belt between Mars and Jupiter.

Galileo weighed 4,989 lb (2,222 kg), and

incorporated a high-gain communications antenna, which unfortunately failed to deploy properly. This reduced the amount of data that could be transmitted to Earth, although engineers found ways to use the alternative systems to their fullest. The main payload on *Galileo* was the 747 lb (339 kg) instrumented descent probe, which was released some 50 million miles (80 million km) from Jupiter in July 1995. On arrival at Jupiter in December it plunged into the cloud deck at a speed of 30 miles per second (47 km/s), surviving for 75 minutes until it was destroyed under the intense pressures.

On its way down it confirmed the existence of an intense radiation belt around the planet at 39,000 miles (50,000 km) distance, and closer to the planet, a cloud layer with wind speeds of 2,000 ft per second (640 m/ps), and traces of organic compounds.

Galileo itself entered orbit on December 8, 1995, beginning an epoch-making exploration of the planet, especially of the Great Red Spot and the four Galilean moons. It will end its career with a suicide plunge into Jupiter's swirling clouds in 2003.

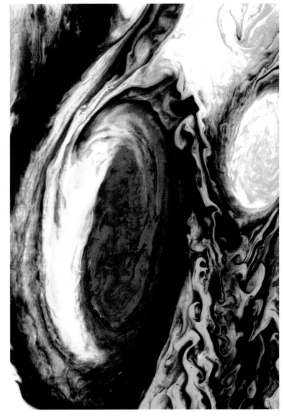

Above and left:

The Voyager and *Galileo* spacecraft returned spectacular images of the swirling clouds, Jovian moons, and the Great Red Spot.

Land of the Giants

Jupiter is certainly a world of superlatives. The Great Red Spot and the four Galilean moons have been photographed by five spacecraft, revealing dramatic vistas. The pictures which these craft sent back to Earth have astounded scientists and the public alike, revealing Jupiter and its moons to be far more dynamic than previously imagined.

Below: The Galilean moon Callisto is heavily cratered and shows little evidence of recent activity, while the icy moon Europa, **center**, has a chaotic, cracked surface.

The famous oval-shaped Great Red Spot in the planet's southern hemisphere had a rather enigmatic personality until spacecraft took a closer look. The craft *Galileo* in particular provided an insight into the dynamics of the swirling hurricane of gases, with speeds that reach 225 mph (360 km/h). In 2002 the Great Red Spot measured 15,000 miles (25,000 km) by 6,850 miles (11,000 km), but has sometimes been up to 30,000 miles (50,000 km) wide. The storm varies in intensity and coloration each year, and moves in relation to the surrounding clouds. Its top towers five miles (8 km) above the rest of Jupiter's clouds and seems to spiral downward like a giant tornado, creating a complex wave pattern. The red color of the Spot reveals a large amount of phosphorous within it.

New moons are continuously being discovered around Jupiter. The number is now well over 20. Four of these were first spotted by Galileo Galilei, and are named after him. The Galilean moons are called Callisto, Europa, Ganymede, and Io. Despite the fact that they are all Jovian moons, they are very different to each other in form.

Callisto has a surface resembling that of an avocado pear, while Europa is like a giant cracked ice ball. This ice surface covers what is thought to be an ocean of water. Ganymede has a cracked, undulating surface. Io is an extraordinary world

with a blistered, volcanic surface. It has sulfur volcanoes, which spew material 185 miles (300 km) out into space.

The 2,980 mile (4,800 km) diameter Callisto orbits 1,170,000 miles (1,883,000 km) from Jupiter. It is the outermost and darkest Galilean moon and the most cratered, showing no evidence of recent geological activity. It has very few craters smaller than 40 miles (60 km) in diameter, and also features a 2,500 mile (4,000 km) diameter impact basin called Valhalla.

Warm ocean on Europa?

The 1,950 miles (3,138 km) diameter Europa orbits 416,700 miles (670,900 km) from Jupiter, close enough to the giant planet to be heated internally by gravity. This gravitational pull creates tidal

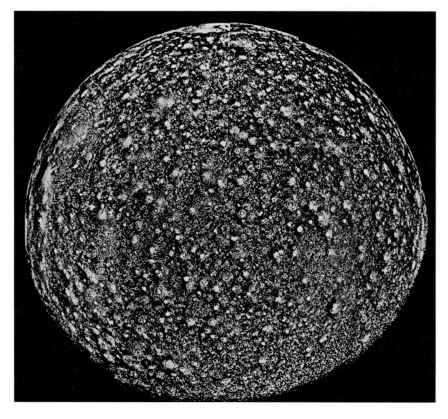

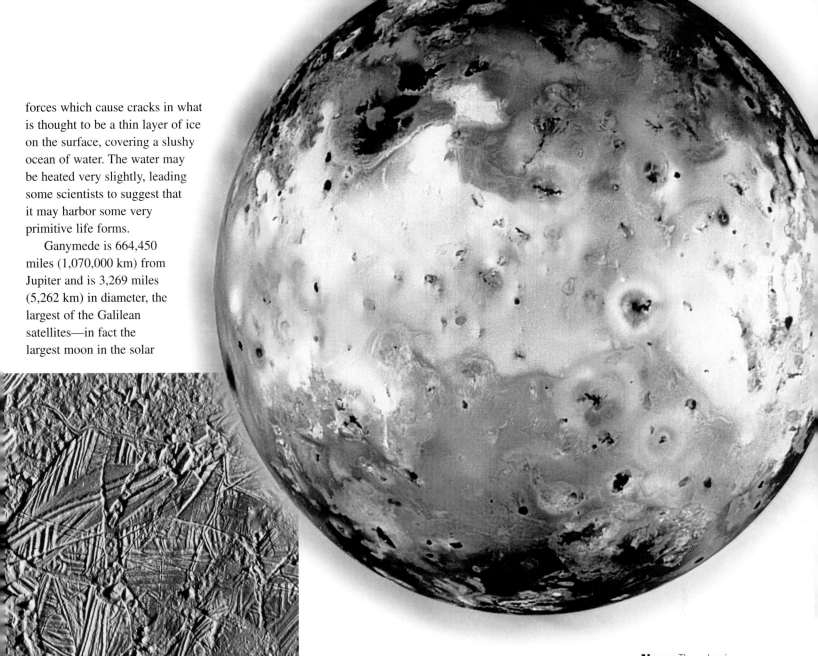

forces which cause cracks in what is thought to be a thin layer of ice on the surface, covering a slushy ocean of water. The water may be heated very slightly, leading some scientists to suggest that it may harbor some very primitive life forms.

Ganymede is 664,450 miles (1,070,000 km) from Jupiter and is 3,269 miles (5,262 km) in diameter, the largest of the Galilean satellites—in fact the largest moon in the solar

system, 8% bigger than the planet Mercury. It is mainly a dirty, icy body comprising water ice with carbon dioxide made dark in color by silicate minerals and tarry molecules. It is quite heavily cratered, with two distinct dark and pale terrains.

The 2,256 miles (3,630 km) diameter Io comes very close to Jupiter in its elliptical orbit around the planet, with an average distance of 261,860 miles (421,600 km). It is distorted by Jupiter's gravity, causing it to bulge out. As it moves away, the moon contracts. This constant movement creates a hell-like surface. The center of the moon heats up and enormous volcanoes erupt, spewing sulfur and sulfurous compounds into space and all over the surface, with shades of orange, yellow, and white. The surface is also covered with molten and solidifying lava.

Above: The volcanic moon Io spews sulfur and compounds out of volcanos, thousands of miles into space, while Ganymede, **left**, is the largest moon in the solar system.

143

Saturn and the "Grand Tour"

The years 1976–80 coincided with a convenient positioning of some of the planets relative to each other in the solar system, which only occurs every 176 years. It was therefore possible to use the gravitational forces of a planet to divert a spacecraft from one to another, like a slingshot, saving time and making a multiple fly-by possible. Pioneer 11 used this technique to move on to Saturn after its Jupiter fly-by, as did Voyager 1 and Voyager 2.

The technique also enabled visits by Voyager 2 to Uranus and Neptune as well (*see pages 148–151*), leaving just Pluto unexplored. Pluto could have been included in what was called the Grand Tour project, but budgets and politics intervened. NASA had ambitious plans to fly Grand Tour gravity-assist missions starting with launches in 1976–77 to Jupiter, Saturn, and Pluto, and in 1979 to Jupiter, Uranus, and Neptune. In 1972, however, only two missions—to Jupiter and Saturn—were approved, under the name Mariner-Jupiter-Saturn, and in 1977 the project's name was changed to Voyager. Thankfully, Voyager 2's flight was extended to include Uranus and Neptune.

Pioneer 11, which was not officially a Grand Tour mission as it was developed in the late 1960s, became the first spacecraft to visit Saturn, on September 1, 1979, passing to within 12,980 miles (20,900 km) of the planet. Its major discovery was that the famous ring system has an extra ring dubbed the F ring. Pioneer 11 also discovered two

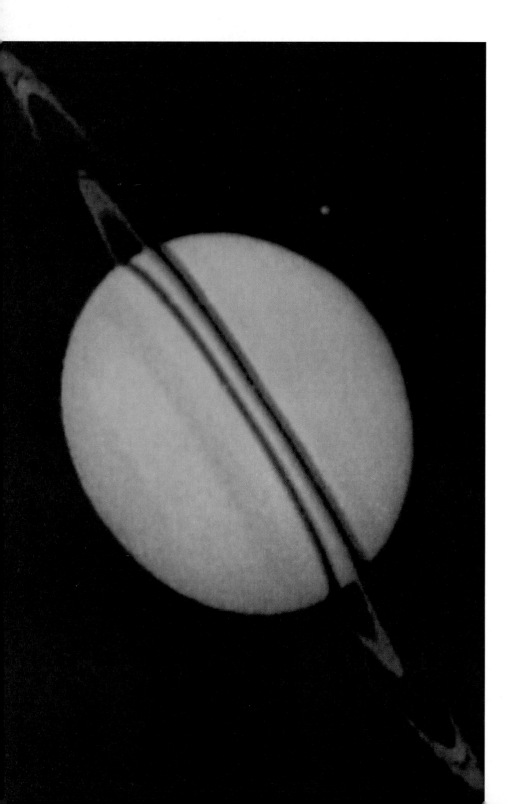

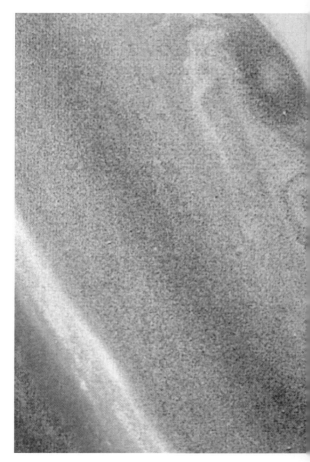

previously unknown moons and found that Saturn's temperature is –180°C (–292°F), and that it is emitting two and half times as much heat as it receives from the sun.

Enigmatic beauty

Through a telescope, the planet Saturn is one of the most beautiful objects in space. Its ring system has fascinated observers for over 300 years. Saturn is the second largest planet in the solar system after Jupiter, with a slightly less turbulent atmosphere of mainly hydrogen. Like Jupiter, because it rotates so quickly (10 hours 40 minutes), it bulges slightly at the equator.

Saturn's maximum distance from the sun in its solar orbit is 936 million miles (1,507 million km) and the minimum, 836 million miles (1,347 million km). It orbits the sun every 29.46 years. The planet is 74,100 miles (119,300 km) across at the equator and 66,890 miles (107,700 km) at the poles. Beneath its hydrogen clouds, Saturn is thought to have a layer of metallic hydrogen, followed by liquid hydrogen and a rocky core in the center.

The first telescopic observation of Saturn was

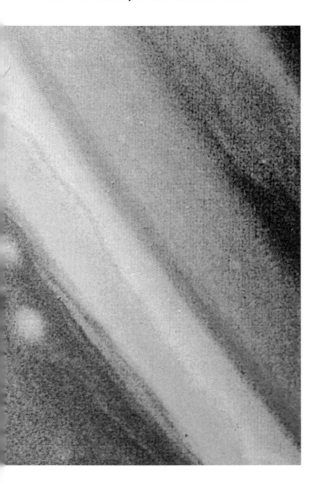

made by Galileo in 1610, but he did not realize he was looking at a ring system despite spotting a bulge around the planet. The first to observe the rings in some detail was Christiaan Huygens, in 1656.

The rings came under close scrutiny during the Voyager fly-bys (*see also Hubble, pages 168–69*), which also added to the knowledge of the dynamics of the planet itself. Voyager 1 arrived on November 12, 1980, passing by 77,000 miles (124,000 km) distant, while Voyager 2 explored the planet from a distance of 63,000 miles (101,000 km) on August 26, 1981.

The rather benign Saturn itself proved to be interesting, but not exactly spectacular, despite the discovery of white, brown, and red ovals in the clouds, rather like poor relations to Jupiter's Great Red Spot. The fly-bys became far more famous for new discoveries about Saturn's ring system and its multifarious moons.

In late 2004, a new spacecraft, *Cassini*, will orbit the planet for the first time and will also deploy a craft called *Huygens*, to land on one of the planet's moons, Titan. *Cassini* was launched on a Titan 4B booster from Cape Canaveral in 1997, and will have made one fly-by of Venus and three of the Earth to set it up for its course to Saturn.

The first view of the ringed planet, Saturn, was taken by Pioneer 11, which made rendezvous in 1979. This was followed by Voyagers 1 and 2 in 1980–81.

Rings and Moons

The Voyager 1 and 2 encounters with Saturn were historic. Many of the planet's moons were revealed in detail, but the biggest discovery was that the famous rings are in fact thousands of separate bands or ring systems orbiting Saturn, held together by the planet's gravitational forces and some "shepherd moons" orbiting in the ring system. The images of Saturn's rings which Voyager 1 transmitted are some of the most spectacular of the Space Age.

T he suspicion that the rings were a more complicated affair than at first thought was confirmed by Pioneer 11's discovery of the F ring. The particles in the ring system range in size from 30 ft (10 m) icebergs to tiny specks less than 0.0005 cm across. The rings have spokes, micron-sized particles levitated electrostatically above the plane of the ring system.

Shepherding moons, later named Prometheus and Pandora, were found and a third and larger moon, called Atlas, 18 miles (30 km) in diameter, was discovered toward the inner part of the system. The gaps in the individual rings are probably locations of other moons waiting to be discovered.

Below: One of the greatest images of the Space Age—a close-up of the rings of Saturn, showing them to have spokes and shepherd moons.

The ring system measures 171,000 miles (275,000 km) across, but is just 30 ft (10 m) high.

Voyager 1 observed all of Saturn's moons that were known at the time. The 242 miles (390 km) diameter Mimas features a crater with a diameter of 80 miles (130 km) that looks like a giant eye. Enceladus features the brightest known surface in the solar system. A huge fracture 465 miles (750 km) wide on Tethys was named Ithaca, while Dione and Rhea display interesting ice features. A cloud of uncharged hydrogen was detected between Rhea and Titan.

Alien worlds

The 3,200 mile (5,150 km) diameter Titan, one of the largest moons in the solar system and the only moon with an atmosphere, proved to be interesting, with a dark red atmosphere totally covering all surface features, and a particularly dark region over its north pole. The surface temperature was measured at -180°C (-292°F), with an atmospheric pressure 1.6 times that of Earth. Titan turned out to have a predominently nitrogen atmosphere rather

than the methane one anticipated by astronomers. However, a resulting atmosphere profile indicated methane clouds at 6–9 miles (10–15 km) above the surface, with argon as another main constituent.

Voyager 2 explored some other moons and took photographs with improved detail of some of Voyager 1's targets. The moon Iapetus displayed a large brown stain on the surface, while Hyperion revealed itself as a 186 mile (300 km) diameter rock that had been in a huge collision with another object. New images of Enceladus revealed it to look rather like Jupiter's Ganymede, while a new feature, a crater with a diameter of 246 miles (400 km), was seen on Tethys.

The stream of new information from Saturn was almost too much to digest, and illustrated the dramatic revolution in astronomy that was taking place. Data captured in just a few minutes during the Voyager fly-bys gave 90% more information than had been gleaned by astronomers over centuries of peering through telescopes.

Left: The Saturnian moon Enceladus reveals a surface like Jupiter's Ganymede, while the montage **below**, depicts the ringed planet and some of its moons, including Mimas in the foreground.

The Train Wreck

Voyager 2 flew past the planet Uranus on January 24, 1986, at a distance of 44,010 miles (71,000 km). The planet was rather benign and a disappointment after the thrills of Saturn, but the moon Miranda saved the day. It turned out to be the great train wreck in the sky.

Uranus was the first planet to be discovered using a telescope, by astronomer Sir William Herschel, in 1781. All the other planets closer to the sun can be seen with the naked eye.

Uranus orbits at a maximum distance from the sun of 1,865 million miles (3,004 million km), and a minimum distance of 1,698 million miles (2,735 million km). The unusual planet has poles that point almost sideways toward the sun, at an angle of 98°. This means that each side of Uranus has 42 years of daylight and 42 years of night, with the north and south poles pointing alternately at the sun, as the planet orbits it every 84 years. The 32,170 mile (51,800 km) diameter Uranus is also unusual in that it rotates in a counter-clockwise direction in 17 hours. It is possible that the planet was knocked over by a large body in the early history of the solar system.

Five moons—Miranda, Ariel, Umbriel, Titania, and Oberon—were originally discovered orbiting Uranus, but many more were found by Voyager 2. The planet has a ring system which was detected by astronomers in 1977. The rings were discovered when the planet passed in front of a star. Astronomers saw that the star seemed to blink a number of times before Uranus actually passed in front of it, indicating that it had a ring system which blocked out the light from the star in brief intervals.

Voyager 2 took close-up images of the thin ring system, which thanks to the spacecraft is now known to comprise of nine rings. The system seems to contain boulders less than 3 ft (1 m) in diameter and to have dust clouds in between, indicating that the ring material may be slowly eroding.

Uranus lacked much detail in its hydrogen-helium atmosphere, but Voyager discovered ten new moons inside the orbits of the known moons, the first named Puck, and two others which act as shepherds in the ring system. These are called Cordelia and Ophelia. Puck is the largest of the inner satellites, with a diameter of 96 miles (150 km), 53,000 miles (86,000 km) distant from Uranus.

A geologist's dream

The moon Oberon displayed a heavily cratered surface with a 3.7 mile (6k m) high mountain and a large trough with scalloped walls. Ariel is cratered, fractured, and the brightest satellite,

Right: A close-up of the thin ring system of Uranus, which comprises small boulders and dust clouds.

featuring valley floors which seem to have been flooded by a viscous fluid as a result of cryovulcanism. Titania has a frost-like appearance with bright craters up to 9 miles (15 km) in diameter and fault scarps 1–3 miles (2–5 km) deep and up to 900 miles (1,500 km) long.

Umbriel is dark and covered in pockmarks. The darkness is thought to be the result of the radiation of methane over many ages. But star of the show was Miranda, which is covered in craters, faults, grooves, terraces, and features a cliff 10 miles (16 km) high, an apparent combination of all the geology in the solar system rolled into one. It looks like a train wreck and appears to have been broken apart and fused together again, maybe a number of times. Many surface features seem to be blanketed in dust or snow, perhaps as a result of cryovolcanism. The three most prominent features are tracts of concentric terrain called coronae, and they were named Arden, Inverness, and Elsinore; names from Shakespeare's plays, and in keeping with the Shakespearian names of many of the Uranian moons.

Above: Voyager 2 bids farewell to Uranus as it embarks on the three-year journey to its next target, Neptune.

Left: The moon Miranda looks as though it has been split apart and the debris fused together again.

Scooter and the Freezer

Neptune was discovered in 1846 by the German astronomer Johann Galle. He was following calculations by other astronomers, who predicted the existence of a planet in Neptune's orbit mathematically. They had noticed that the orbit of Uranus was being affected by an unknown force of gravity, which they assumed was another planet.

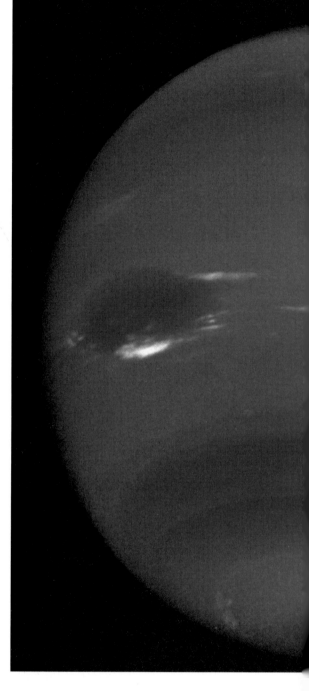

Voyager 2 flew past Neptune on August 24, 1989, at a speed of 16.8 miles per second (27 km/s) and at a distance of 3,000 miles (4,900 km), ending its great exploring career. The craft found that Neptune is a more vibrant planet than the rather subdued Uranus, and features high altitude clouds in a gaseous hydrogen and helium atmosphere that scoot around the planet faster than it rotates, in 19 hours. The biggest, called Scooter, traveled at over 1,490 mph (2,400 km/h). A Great Dark Spot the size of the Earth was also found, but this feature and Scooter were not seen in more recent images taken by the Hubble Space Telescope, so presumably were transient in nature.

It was also discovered that Neptune has a small ring system. Although discovered by Voyager 2, the

Below: Voyager 2's views of the coldest place in the solar system—the Neptunian moon, Triton.

ring system was suspected in 1846. Voyager also found six more moons. Of the two previously-known moons, the largest, Triton, has one of the most dynamic surfaces in the solar system. At -235°C (-391°F), it is the coldest known place in the solar system. The 1,860 mile (3,000 km) diameter Triton is a chaotic world of cliffs, craters, mountains, glaciers, and liquid nitrogen geysers. These are evidence of cryovolcanism, where viscous melted material has oozed out, in one place creating long ridges called cantaloupe terrain, pockmarked

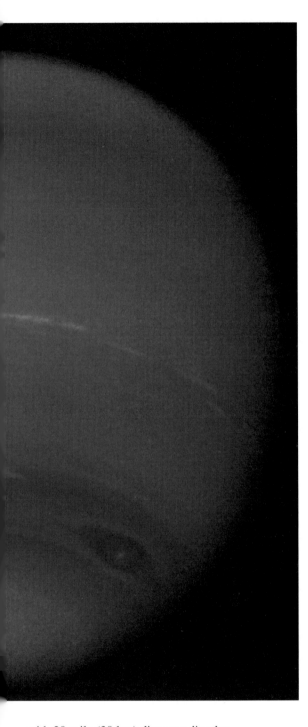

Cosmic clash

Another moon, Neried, 310 miles (500 km) in diameter, is in a very deep 800,000 mile (1.3 million km) by 6 million mile (9.76 million km) orbit in the opposite direction. These unusual orbits may have been caused by other moons which strayed too close, throwing Triton and Neried out of their paths and themselves out of Neptune's orbit entirely. Those moons may have been the furthest planet Pluto and its companion Charon. Pluto's orbit periodically reaches inside Neptune's.

The 30,000 mile (49,500 km) diameter Neptune orbits the sun at a maximum distance of 2,810 million miles (4,537 million km), and a minimum distance of 2,767 million miles (4,456 million km), and takes 165 years to complete an orbit of the sun. Its distinctive blue color results from the hydrogen-helium atmosphere. Cirrus clouds of frozen methane cast shadows on Neptune's blue atmosphere 30 miles (50 km) below. Beneath the hydrogen and helium clouds lies a possible ocean of water, ammonia, and methane, with a rocky core at the center.

Voyager's exploration of Neptune ended a dramatic phase in planetary exploration that may never be repeated. In the space of just 27 years, spacecraft explored every planet in the solar system first-hand, except for Pluto.

Left: The full view of Neptune taken by Voyager 2 shows the scooter clouds moving rapidly across the atmosphere, faster than the planet rotates.

Below: A final view of Neptune as Voyager 2 ends its epic odyssey in planetary exploration.

with 20 mile (30 km) diameter dimples.

Voyager spotted several jets of gas bursting through the south polar cap as a result of internal heating, and projecting columns of sooty particles 5 miles (8 km) high, at which point they are deflected sideways by high altitude winds. Triton has a circular orbit moving east to west in the opposite direction to the planet's rotation. The retrograde orbit, at an angle of 157° to Neptune's equator, makes it unique among all the solar system's moons.

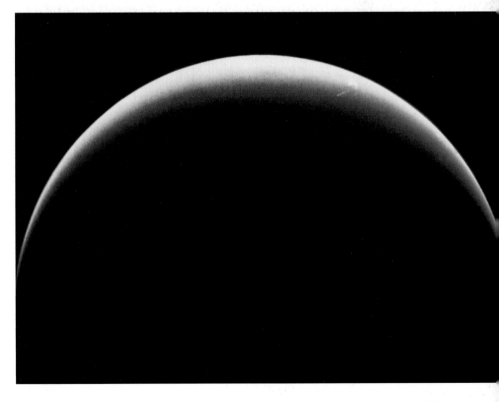

Flying to the Rocks

The first known asteroid fly-by took place on October 29, 1991. En route to Jupiter, *Galileo* flew to within 1,000 miles (1,600 km) of the 11 mile (18 km) long main-belt asteroid 951 Gaspra. It looked like a large irregular-shaped pockmarked pebble with craters. The appearance of an asteroid was not that surprising since the Martian moons, Phobos and Deimos, assumed to be captured asteroids, had already been explored.

Right: The Near Earth Asteroid Rendezvous spacecraft is launched aboard a Delta II from Cape Canaveral in 1997.

Galileo's encounter was followed by a 1,500 mile (2,410 km) fly-by of the 32 mile (52 km) long Ida on August 28, 1993, which turned out to have a 1 mile (1.6 km) "moon," called Dactyl.

In the 1990s, NASA introduced a new class of technology demonstration spacecraft in its New Millennium program, and the first of these was called Deep Space 1, which incorporated an ion propulsion system and new microelectronic instruments. Deep Space 1 flew by asteroid Braille in July 1999, at a distance of just 16 miles (27 km), but the imaging system malfunctioned. A bonus fly-by of the comet 19P/Borrelly was made in September 2001.

The first dedicated asteroid explorer was the Near Earth Asteroid Rendezvous (NEAR) spacecraft, flying the first NASA Discovery "faster, better, cheaper" series mission, to explore the asteroid Eros. NEAR was launched in February 1997, and in June flew close by the asteroid Mathilde, aiming to enter orbit around Eros by February 1999. Unfortunately, computer and engine errors nearly aborted the mission, but NEAR was rescued and made it safely to Eros a year later, the first spacecraft to orbit an asteroid.

NEAR entered an initial, 228 by 120 miles (366 km by 200 km) orbit, and for a year the craft was maneuvered to fly past Eros at altitudes ranging from 3–35 miles (5–56 km), producing thousands of spectacular images, showing large and small craters, valleys, and boulders. Initial images showed evidence of a layered structure, which may indicate that the 20.5 mile (33 km) diameter asteroid is a remnant of a larger parent body which broke apart. Also, the asteroid has a higher density of craters than observed during fly-bys of main-belt asteroids.

Combining digital images and data from the craft's laser rangefinder, scientists built the first detailed map and three-dimensional model of an asteroid. Data suggests that Eros is a cracked but solid rock, made of

Left: *Galileo's* view of the asteroid Ida was taken in 1993. The 32 mile (52km)-long rock proved to have a small satellite, Dactyl.

Below: NEAR's view of Eros, on which the spacecraft would make an historic soft-landing on February 12, 2001.

some of the most primitive materials in the solar system. The regolith on Eros was found to be about 300 ft (90 m) deep in places, smoothing over rough areas and spilling into craters.

The cratering on Eros surprised scientists, with intriguing square ones and many fewer small craters than expected. More than 100,000 craters wider than 50 ft (15 m) were counted. The one million or more boulders seen on the surface were totally unexpected. Some were the size of a house or even larger.

Historic landing

Toward the end of 2000, the orbit of NEAR *Shoemaker* (the craft was renamed in memory of the famed astrogeologist, Gene Shoemaker) was again reduced with the idea of enabling the craft to fly to within a few meters of the surface at very slow speed, before crashing. However, rather than do this, NASA decided that it would try to make an unprecedented soft-landing. So, on February 12, 2001, NEAR *Shoemaker* gently landed on the tips of its two solar panels with its bottom edge on the surface. History was made.

The spacecraft took 69 detailed pictures during the final 3 miles (5 km) of its descent, showing features as small as one centimeter across. The slow touchdown speed left the spacecraft intact and still sending a signal back to Earth, giving "bonus science." The transmissions continued until February 28, when the spacecraft died.

Exploring the Snowballs

Comets are believed to be composed of the primordial material from which the solar system formed, and as such are of great interest to scientists. In 1986, the much-heralded return of Halley's Comet occurred, and for the first time in its regular 76-year visits, it was explored by an international fleet of spacecraft from Russia, Japan, the U.S., and Europe. Since then, several other comets have also been explored.

Below left: Britain's *Giotto* spacecraft made one of the most difficult and historic voyages in space history—through the coma of a comet, in 1986.

Below right: *Giotto* 's close-up of the nucleus of Halley's Comet spewing material from its highly active surface, as it passes close to the sun in its 76-year solar orbit.

The prime craft at Halley's Comet was Europe's *Giotto*, designed to fly right through the coma of the comet in one of the most remarkable, and yet largely unheralded, flights in history. *Giotto* was a spin-stabilized, drum-shaped craft built in Britain. It weighed 2,116 lb (960 kg) at launch and was 16.12 ft (8.67 m) in diameter, and 9.33 ft (2.8 4m) high.

It was protected by a shield of aluminum and Kevlar from impacts of comet dust and particles as the craft traveled through the comet at a speed of 42 miles per second (68 km/s)—fast enough to make a trans-Atlantic trip in just over one minute. The remarkable *Giotto* shot through the top part of the coma of Halley's Comet on March 13, and

found that ten tons of water molecules and three tons of dust were being thrown out of the comet every second, as its shield took a battering, which could be heard in transmissions to Earth.

The best image, from 11,250 miles (18,000 km) away, showed that the nucleus is 9 miles (15 km) long and between 4–6 miles (7–10 km) wide, with two large jets of dust and gas erupting from what appear to be cracks in the undulating surface of hills and valleys. Further data was returned from the other spacecraft, the Soviet Union's Vega 1 and 2, and Japan's Sakigake and Suisei, from a greater distance away.

Fourteen years later, on September 22, 2001, NASA's Deep Space 1 flew past comet 19P/Borrelly, but images were not clear due to a camera fault. The comet fly-by was not originally planned for the technology demonstration mission but was added as part of a trouble-shooting procedure.

Sample-return mission

The next comet explorer will be *Stardust*—in fact it will bring back dust from the comet Wild-2 inside a special re-entry capsule in 2006. The NASA Discoverer mission series spacecraft was launched on February 7, 1999. As its name implies, *Stardust* will also spend 150 days collecting particles of interstellar dust, including recently discovered grains of matter streaming into the solar system

from the direction of Sagittarius. This dust consists of ancient interstellar matter dating from before the formation of our solar system.

In February, 2000, *Stardust* successfully deployed an aerogel collector and began collecting interstellar dust. Aerogel is a silica-based solid with a porous, sponge-like structure in which 99% of the volume is empty. This historic first collection period lasted through to May 1. The collector was then returned to its stowed position until mid-2002, when another period of interstellar dust collection was scheduled.

For the first time ever, comet dust will be collected, during the close encounter with Comet Wild-2 in January 2004. The objective is to recover more than one thousand particles larger than 15

microns in diameter. *Stardust* should come within 93 miles (150 km) of the comet at its most active, as it approaches the sun, and will also send back images and count the number of comet particles striking the spacecraft. Real-time analyses of the compositions of the particles will also be attempted.

After *Stardust* has collected the comet dust samples, all the particles captured in the aerogel collector will be retracted into the sample return re-entry capsule as the craft flies close to the Earth again in 2006. The samples will then be returned to Earth via parachute for a soft landing at the U.S. Air Force's Utah Test and Training Range.

Another Discovery spacecraft, the Comet Nucleus Tour (Contour) was launched successfully on July 2, 2002, en route to two comets, but broke apart in space a month later.

Dreams and Reality

The Space Age is now almost 50 years old. Perhaps the main difference between the attitude today and in the 1950s is a sense of reality about what can be accomplished, at what cost, and in what time-frame. Our exploration of the solar system using unmanned probes has been extremely successful, but experience has also shown that manned missions to the planets are still a very long way off.

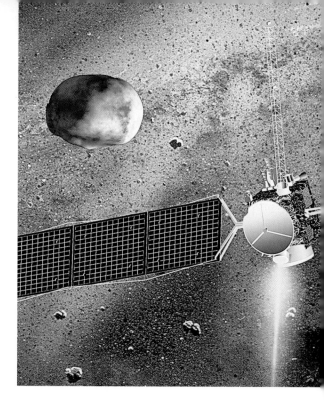

We can have dreams of colonizing the moon and Mars, but that is all they are. However, many future unmanned planetary exploration projects have been given the go-ahead, and these should be separated from the still-theoretical missions that have been proposed.

For example, many NASA scientists want to send a spacecraft to orbit one of Jupiter's moons, Europa (*see page 142*), and then land a craft on the surface and penetrate its ice layer with a cryobot, which will search for signs of life in the waters. This mission has not been approved, however, and would take many years to accomplish, as would a proposed NASA mission to Pluto and the Kuiper Belt of objects in the distant solar system. Even if the Pluto spacecraft is launched soon, it would take over a decade to reach the planet. If the mission is

Below: A planned NASA flight to the most distant planet, Pluto, and its large moon Charon, has not yet been fully funded.

to happen at all, it will have to be launched soon, since in a few decades Pluto will be completely frozen over and unexplorable. Its budget has not been cleared.

Budgets for missions such as these are a minimal few millions, unlike the days of the multi-billion dollar Voyager program which was launched in the 1970s. NASA's Discovery program, with the tagline "faster, better, cheaper," includes several firm future planetary missions, including the Mercury Surface Space Environment Geochemistry and Ranging mission, Messenger (*see page 120*). Europe plans the Mercury Explorer Bepi Columbo, and there are proposals to launch further Mars missions, including the sample return mission (*see page 138*).

The next Discovery program mission, Deep Impact, will be launched in January 2004 to rendezvous with the comet P/Tempel 1 in July 2005, flying by at a distance of 310 miles (500 km). The craft will eject a 770 lb (350 kg) copper projectile traveling at 20,000 mph (32,200 km/h) into the nucleus of the comet to measure its composition, while the main craft observes the action.

Slow journey times

Two asteroids, Ceres and Vesta, the first and largest to be discovered, will be observed at close quarters by *Dawn*, powered by an ion propulsion system. It will be launched in May 2006, and will reach Vesta in 2010 and Ceres in 2014. *Dawn* will be powered by an ion propulsion system demonstrated successfully by the NASA New Millennium project

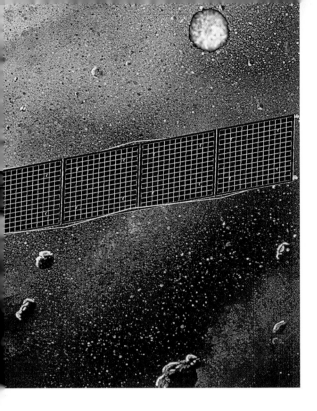

spacecraft Deep Space 1.

A major European Space Agency project is the international Rosetta flight to explore two asteroids and the comet Wirtanen, in a mission which will begin in January 2003 but which will not reach its prime destination, Wirtanen, until 2011. The two asteroids, Otawara and Siwa, will be explored in 2006 and 2008, while the flight will also involve Earth gravity-assist swing-bys in 2005 and 2007.

Rosetta will deposit a lander onto Wirtanen's surface, to measure the composition and structure of the nucleus' material.

Bepi Columbo will be powered by an ion propulsion system, to be demonstrated by another ESA craft, Smart 1, which will orbit the moon in 2003, while Japan plans to launch a Lunar-A orbiter mission to send penetrators into the moon's surface equipped with seismometers and heat-flow probes, one on the near side and one on the far side. The orbiter will also deploy two sub-satellites to act as communications relays for the penetrators. Lunar-A is billed as the biggest mission to the moon since Apollo. Japan also plans another mission, Muses-C, to explore an asteroid and bring some material to Earth. All these missions depend on budgets being finally cleared. Lunar-A has already been delayed several years.

India and China have said that they plan to launch probes to the moon, but no budgets or timelines have been decided.

Left: The USA's *Dawn* spacecraft will be launched in 2006 to explore two of the largest asteroids, Ceres and Vesta, in 2010–14.

Below: The European Space Agency's *Rosetta* spacecraft will reach the nucleus of the comet Wirtanen in 2011.

CHAPTER ELEVEN

THE SPACE BUSINESS

Space Exploitation

Most people in the industrial world are oblivious to the fact that satellites are providing many of the vital services that support their daily lives. Developing countries are also benefiting from space today. Over the past 40 years, global society has come to rely on satellites for uses ranging from mobile phone networks to global positioning system (GPS) navigation.

Sixty-six satellites and spacecraft were launched in 2001. There were 58 launches, with some rockets carrying multiple payloads. Of these, 23 were made by the U.S. and 23 by Russia, while there were eight launches made by Europe, two by India, and one each by China and Japan.

Of the 66 spacecraft, 15 were related to manned spaceflight, 13 were military, five were for Earth observation, four for GPS navigation, three for science, and two each were technology and planetary missions. The majority of the remaining 22 were communications satellites, mostly those which operate in geostationary orbit. Communications satellites are providing many services, including international mobile phone calls and, of course, satellite TV, not to mention playing a major role in the rapid growth of the Internet. A typical satellite in geostationary orbit is the size of a house, with a wingspan of 200 ft (60 m), and equipped with 20 C-band and 40 Ku-band transponders. The C-band transponders are used to serve cable TV customers, while the Ku-band payload is used for video distribution, data networks, and broadband service for the Internet.

Growing range of uses

Remote sensing Earth observation satellites are providing services to a variety of industries and users, including geologists, urban planners, and environmentalists. Remote sensing data is transformed into image products, often called Geographical Information Systems (GIS), combining the information from several types of image of the same area, and data from other sources, such as ground surveys and maps.

Aircraft, ships, and vehicles navigate using the Navstar Global Positioning System (GPS) satellites, and road fleet management is assisted by data messaging and positioning satellites. This is supported by Russian Glonass satellites, and soon a European network of Galileo navigation satellites may be launched.

Many satellites are in view to users in most parts of the populated world at any time, and signals can calculate time, location, and velocity

to an accuracy of a millionth of a second, a fraction of a mile per hour, and to within a few feet of any location. GPS receiver units are onboard aircraft, ships, land vehicles, on man-packs, and in hand-held units. In the civilian sector, new applications are continually emerging, but the accuracy is more limited than for those in military service. Daily TV weather forecasts feature satellite images from spacecraft, while other satellites monitor the Earth's environment, providing such data as wave height and sea temperature.

The leading space nations of the world operate a co-operative fleet of polar-orbiting and geostationary satellites, which cover the whole globe, returning meteorological and environmental data 24 hours a day, ranging from the visible wavelength images of the Earth we sometimes see on TV, to surface temperature maps.

Military operations, such as anti-terrorist actions, are supported by a fleet of satellites providing services ranging from communications and reconnaissance to missile early-warning.

Fleets of different kinds of communications satellites enable Naval fleets, aircraft, and ground troops to communicate, and to receive the latest intelligence data from other satellites. Spacecraft take digital images relayed directly—or via data relay satellites—to the U.S. National Reconnaissance Office in Washington, and these can also be sent to any battlefield. Ocean reconnaissance satellites monitor fleet movements.

Electronic intelligence satellites, called Elints, such as the CIA's Magnum, monitor radio transmissions and radar emissions from military installations—and can even be used to bug civilian telephone calls. Some elints are like giant vacuum cleaners, with a receiver dish up to 327 ft (100 m) in diameter, which collect transmissions from many sources. The data can then be unscrambled in ground centers.

Above: A heat detecting Department of Defense early warning satellite monitors missile launches and can even spot a jet on afterburner.

Left: A European geostationary-orbiting Meteosat provides a 24-hour watch on the weather and environment over one third of the Earth.

Opposite: A Tracking and Data Relay Satellite system operated by NASA as part of a network which enables the Space Shuttle and International Space Station to remain in almost constant touch with ground control.

159

The Space Workhorses

Satellite launchers today are developed to serve a market, rather than the market being dependent on the size of the launcher as was previously the case. Another transformation has been a bias away from government funding to privatization of launch operations.

Below right: Europe's Ariane 5 is operated at Kourou, French Guiana.

Below left: The new Lockheed Martin Atlas V booster, pictured at Cape Canaveral, made its maiden flight in 2002.

A major turning point was the *Challenger* accident in 1986. The Shuttle had monopolized the U.S. launcher business, and many unmanned vehicles were on the point of being scrapped. The U.S. had made a very risky decision in having just one launch system, and that decision backfired. The Space Shuttle was grounded, and commercial business launches ceased.

Several American launchers were spared as a consequence, and a vibrant private commercial launcher industry developed, with additional competition from other countries, such as European nations and Russia after the collapse of the Soviet Union. In the past few years, co-operation between the U.S. and Russian space industries has reached such an extent that the latest U.S. launcher, the Atlas V, is powered by a Russian rocket engine.

Some of the major launchers still have a ballistic missile heritage. The Russian ICBM which launched Sputnik 1 gradually became the Soyuz. The American Atlas ICBM gave birth to the new Atlas V. Today's Delta II is a direct descendent of the Thor IRBM. The Titan II second generation ICBM still flies today as the core stage of the Titan IVB-Centaur. With the introduction of Delta IV and Atlas V fleets by the U.S. Air Force, which will also fly commercial missions, the Titan will finally be retired.

The commercial market is led by Arianespace, a European consortium, which operates a fleet of highly successful Ariane 4 and 5 rockets from Kourou in South America. The first Ariane flew in 1978, and today its successor, the Ariane 4, is offered in a variety of combinations, with a mix of

liquid and solid propellant strap-on boosters. The new, more powerful Ariane 5 has two large solid rocket boosters. Four high-power models are planned for this booster, which will take over from the Ariane 4 totally in 2003. Arianespace's order book usually numbers about 40 satellites at any time.

International competition

An Atlas fleet is operated from Cape Canaveral, Florida by ILS (International Launch Services). It also offers the Russian Proton booster, which is launched from Baikonur. A new Russian vehicle called Angara may be introduced to replace the Proton. Competing with the Atlas is the Delta fleet, also operated out of Cape Canaveral. The new Atlas V and Delta IV boosters will compete head-to-head with Ariane in a market that has been rationalized by a business downturn.

Russia also operates a fleet of other launchers, including the Zenit 2 and the Cosmos. A former Russian missile, the SS-19, has been converted into the Rokot satellite launcher, marketed by a Russian-German company called Eurokot, which operates out of Plesetsk, Russia.

The international Sea Launch venture operates the Ukrainian-built Zenit 3SL from Odyssey, a semi-submersible launch platform that is floated into position on the equator for accurate launches into geostationary orbit. Satellites have also been launched from the air and from under the sea. The U.S. Pegasus satellite launchers are winged rockets released from underneath an carrier aircraft. A former Russian military missile, the Shtil 2, is fired from a submarine in the Barents Sea.

China launched its first satellite in 1970 using a modified ICBM called the Long March 1, and today offers a range of boosters, including the 3B, which can compete with the Atlas and Delta. Japan has launched several satellites using a series of boosters, leading to the development of the H2A, which operates from Tanegashima, while it also launches smaller boosters from Kagoshima.

Other countries have developed satellite launchers, including Brazil, India, and Israel. Brazil has twice tried to launch a satellite using its small VLS booster, but both launch attempts failed. India operates a Polar Satellite Launch Vehicle and the Geostationary Satellite Launch Vehicle, while Israel has a Shavit booster based on a military missile. North Korea is reported to be planning to launch a satellite using a version of its Tapeo Dong 2 missile.

A Russian Proton booster is launched on a commercial mission from the Baikonur Cosmodrome in Kazakhstan.

Everyday Space Science

As the Hubble Space Telescope and other astronomical satellites receive public attention, many other unheralded science satellites are hard at work. Many are monitoring the Earth's space environment, such as radiation belts, the upper atmosphere, and the ozone layer. These carry out the everyday scientific monitoring and observation of our planet.

The Upper Atmosphere Research Satellite (UARS) was deployed into Earth orbit by the Space Shuttle in September 1991. UARS was the first NASA "Mission to Planet Earth" program spacecraft, and the largest ever flown for atmospheric research. It investigated the processes controlling the structure and variability of the upper atmosphere, to help create a comprehensive

carrying four instruments to study the creation and content of aurorae. Plasma and electron temperature data is collected by electric field detectors, while magnetic fields are monitored by a magnetometer, and electrons and ions by electrostatic analyzers.

Japan's Geotail satellite, launched in 1992, measures global energy flow and transformation in the Earth's magnetotail, to increase our understanding of the processes in the magnetosphere. The craft used two gravity-assist fly-bys of the moon to enable it to explorer the distant part of the magnetotail, while two years later, the orbit was reduced so that the spacecraft could study the near-Earth magnetotail processes.

Right: NASA's Upper Atmosphere Research Satellite was the first "Mission to Planet Earth" spacecraft.

database required to better understand the depletion of ozone in the stratosphere. UARS helped confirm that man-made chlorofluorocarbons are responsible for ozone depletion.

The NASA satellite, Fast Auroral Snapshot Explorer, FAST, was launched in August 1996,

Understanding Earth's environment

NASA's Imager for Magnetopause to Aurora Global Exploration, IMAGE, was launched into a polar Earth orbit in March 2000. It uses neutral atom, ultraviolet, and radio imaging techniques to study the magnetic phenomena involved in the

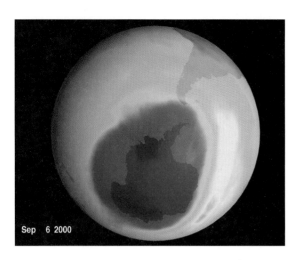

Sep 6 2000

interaction of the solar wind with the Earth. Mission objectives are to identify the main mechanisms which inject plasma into the Earth's magnetosphere, determine the response of the magnetosphere to changes in the solar wind, and discover how and where magnetospheric plasmas are energized and transported.

NASA's Polar was launched in February 1996, equipped with 11 instruments to provide complete coverage of the inner magnetosphere and obtain global images of the polar aurorae. Polar also measures high latitude entry of solar wind and ionospheric plasma, and the deposition of energy into the upper atmosphere. The satellite was designed to have on-board instrument-communications interconnectivity to share data efficiently among the instruments using onboard computers.

A U.S. satellite called Wind was launched in November 1994 to work with the Polar satellite. Wind investigates sources, acceleration mechanisms, and propagation processes of energetic particles and the solar wind. It will help to complete research on plasma, energetic particle, and magnetic field influences on the magnetosphere and ionosphere. Wind was placed into an L-1 halo orbit between the Earth and sun via a lunar gravity-assist swing-by. It is equipped with a suite of ten instruments provided by the U.S., Russia, and France. The Konus instrument was the first instrument from the former Soviet Union to fly on a U.S. spacecraft.

The U.S. Total Ozone Mapping Spectrometer (TOMS) was launched in July 1996. TOMS was dedicated to mapping global atmospheric ozone levels, the key to understanding ozone depletion. The spacecraft complements other craft and

instruments on meteorological satellites, keeping a close watch on the status of the hole in the ozone layer.

A NASA spacecraft, Gravity Probe, will be launched in about 2003 to test Einstein's theory of general relativity by measuring the precession of gyroscopes in Earth orbit. The spacecraft will slowly roll about the line of sight to a guide star viewed by a reference telescope, while four gyroscopes, cooled by liquid helium, will compare data from the telescope to an extreme accuracy. The polar orbiting satellite is to look for the expected 6.6 arcsec/yr precession to the Earth's rotational axis and the 0.042 arcsec/yr parallel to it.

Left: The Total Ozone Mapping Spectrometer spacecraft monitors the state of the hole in the ozone layer.

Below: The Gravity Probe B will test Einstein's theory of relativity using a liquid helium-cooled telescope.

CHAPTER TWELVE

HUBBLE

Bringing the Universe Home

The Space Shuttle was designed in the 1970s to carry a multitude of payloads, including a revolutionary astronomical telescope that would be able to peer 50 times deeper into space than the most powerful Earth-bound telescope. The space telescope was named after Edwin Hubble, the famous astronomer.

Probably the greatest observational astronomer of the 20th century, Edwin P. Hubble (1889–1953) was staff astronomer at the Carnegie Institute's Mount Wilson Observatory, California, from 1919 until his death in 1953.

In the 1920s and 1930s, Edwin Hubble analyzed the speeds of recession of a number of galaxies, and showed that the speed at which a galaxy moves away from us is proportional to its distance. This is now called Hubble's Law. The simplest explanation of this recession is that the universe began with a Big Bang, which is estimated to have happened somewhere between 10 and 15 billion years ago.

The light from an object 10 billion light years away has, of course, taken 10 billion years to reach Earth. By looking at the most distant objects, astronomers are therefore seeing the oldest objects in the universe. The Hubble telescope was designed to view objects 14 billion light years away, in order to see the primordial universe, including, for example, galaxies in the making. It was hoped that by doing this, the telescope would discover how the universe was formed and how old it really is. It would also be able to take photographs of objects closer to home, such as the planets in our own solar system, at a much higher resolution and quality than is possible using Earth-bound telescopes.

The Hubble Space Telescope (HST) was to have been launched on the Space Shuttle in the mid-1980s, but did not fly until 1990. By 2002 it was still operating, with an estimated five more years to go before it would be succeeded by a new generation telescope.

Astounding images

The HST views of the universe have captivated the world more than any other current space project. Although the HST's resolution has now almost been matched by new ground-based telescopes, such as the European Space Observatory, its range of sensors and wide field still contribute to its

popularity—and of course to the great advances in astronomy that have been made in the last two decades.

The HST was designed to be serviced in orbit by Space Shuttle crews, and to have instruments removed and replaced, including its twin, 40 ft (12.19 m) long solar arrays. The 25,500 lb (11,600 kg) telescope is 42.5 ft (13 m) long, and up to 14 ft (4.2 m) in diameter.

It is equipped with two high gain antennas, which enable it to transmit direct to the ground and to use a Tracking and Data Relay Satellite system. It has two low-gain antennas, a data management system, including a high-power computer; and fine-pointing system which enables the HST to remain locked on any specific target to within 0.01arcsec.

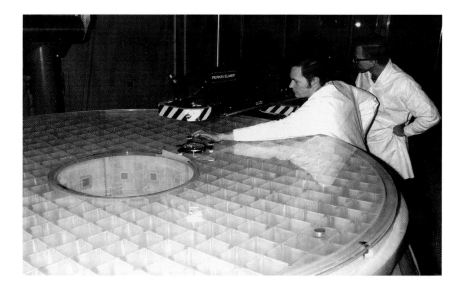

Unfortunately, when Hubble reached orbit, it soon became clear that the telescope had a big problem, since the first images returned to Earth were extremely fuzzy. It was deduced that the primary mirror was not as perfectly curved as anticipated, and that the mirror was suffering from a spherical aberration, which it is said occurred during manufacture. HST needed "a pair of glasses." A team of technicians was launched on the Space Shuttle in 1993, in the hope of being able to repair the telescope. Their chances of success were not good.

Above: The highly-polished primary mirror of the Hubble Space Telescope before assembly of the spacecraft.

Left: An early concept of the Space Telescope.

If HST was in Los Angeles, it could focus on a dime in San Francisco.

The Optical Telescope Assembly is configured in such a way that the telescope is the equivalent of a 189 ft (57.6 m) long telescope, but is in fact compacted to 21 ft (6.4 m). Light entering the aperture door travels down the tube onto a 94.5 in (2.4 m) primary mirror, and is reflected onto a secondary mirror, 12.2 in (0.3 m) in diameter, and then through a hole in the center of the primary mirror onto the focal plane. HST's various science instruments then receive the light.

HST was originally equipped with a Faint Object Camera, Wide Field/Planetary Camera (WFPC), Goddard High-Resolution Spectrograph (GHRS), Faint-Object Spectrograph (FOS), High Speed Photometer (HSP), and Fine Guidance Sensors (FGS).

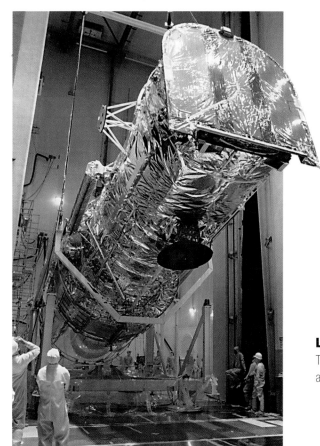

Left: The Hubble Space Telescope is completed and is ready for launch.

From Myopia to Magnificence

The Hubble Space Telescope (HST) was launched aboard the Space Shuttle STS-31/*Discovery* on April 24, 1990, and deployed into orbit using the orbiter's remote manipulator system (RMS). The telescope operates in a 330-mile (530km) circular orbit, inclined to the equator at 28.5°. Following the discovery of the myopia problem, the only way to carry out repairs on Hubble was to join it in orbit and work on the telescope during a series of spacewalks.

The spherical aberration problem could have been exacerbated by minute structural damage to the Hubble's lightweight framework by dynamic launch loads, which were still not fully understood by NASA (*see page 91*).

The "glasses" were called the Corrective Optics Space Telescope Axial Replacement unit (COSTAR), which is about the size of a phone booth. COSTAR was flown to the HST by Space Shuttle STS-61 on December 2, 1993.

During a stunning mission, led by four spacewalkers; Story Musgrave, Jeff Hoffman, Tom Akers, and Kathryn Thornton, COSTAR was fixed inside the telescope. The spacewalkers also replaced a solar panel, repaired electronics, installed a new computer processor, replaced the WFPC with a new unit, installed magnetometers, removed the HSP, and installed a redundancy kit for the GHRS in the most impressive example of EVA work carried out in space to date.

The result of the installation of COSTAR was startling. The HST images were now truly spectacular and the telescope immediately caught the imagination of the public worldwide.

Regular maintenance

After four years providing superb images and data, another HST servicing mission by Shuttle astronauts was launched on February 11, 1997. The STS-82/*Discovery* astronaut crew made a comprehensive change to Hubble's suite of instruments and conducted routine spacewalk servicing. The GHRS and FOS were removed and replaced with a Space Telescope Imaging Spectrograph and a combined Near-Infrared Camera and Multi-Object Spectrometer.

The astronauts replaced an FGS and other equipment including tape recorders, installed an optical electronics enhancement kit, and changed the HST stabilization and fine pointing reaction wheel assemblies. Other

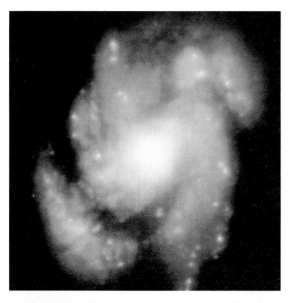

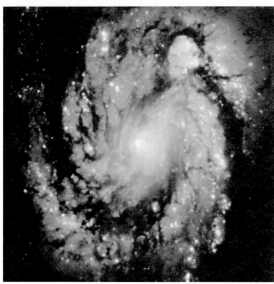

The 3B mission was launched as STS-109/*Columbia* on March 1, 2002, to install an Advanced Camera for Surveys, new rigid solar panels, a new power unit, a new cryo-cooler, and a reaction wheel assembly. A fourth mission is planned for 2003 carrying a Cosmic Origins Spectrograph and a third replacement WFPC.

The Next Generation Space Telescope is planned to succeed Hubble, but will not be launched for several more years, so the HST will remain operational for as long as possible. Another Shuttle mission is being discussed as a grand finale—to bring the HST back to Earth and display it at the Smithsonian Air and Space Museum in Washington, DC.

Opposite: The Hubble Space Telescope was serviced by Space Shuttle astronauts and the corrective optics installed in 1993.

Far Left: The M100 galaxy in the Virgo cluster taken before and after the installation of the COSTAR.

Left: The Hubble Space Telescope pictured after a later Shuttle mission to install new solar panels.

work involved laying new thermal insulation blankets.

Hubble worked on, entering its tenth year of service when a new Shuttle servicing mission was launched. This was originally planned for 2000, but when gyroscopes on the telescope failed at a critical level, the mission was brought forward to December 1999 and split into two, with the second half of the mission being shifted to 2001. STS-103/*Discovery* was launched on December 19, 1999, and the 3A mission installed six new gyros and voltage/temperature kits, a new computer 20 times faster and with six times more memory, a new digital tape recorder, and replaced one FGS and a radio transmitter. The spacewalking astronauts also placed some new insulation over part of the HST.

Bringing the Planets Closer to Home

Saturn

The ringed planet Saturn orbits the sun every 29 years. It has an axial tilt of 26° so that the Earth's view of the planet changes. The planet's rings therefore provide an excellent visual illustration of the tilt, as the view of the rings varies from year to year. This montage of Hubble images shows Saturn between 1996 and 2000, with the rings opened up almost fully, and just past edge-on. The rings are extraordinarily thin, about 30 ft (10 m) through, and are made of dusty water ice in the form of various-sized chunks, which gently collide with each other. Saturn's gravitational field constantly disrupts these chunks, and according to scientists, this keeps them spread out and prevents them from combining and forming a moon. It is believed that the pale reddish color of the rings is due to the presence of organic material mixed with the water ice.

Mars

In June 2001, the planet Mars was just 43 million miles (68 million km) from the Earth, its closest approach to our planet since 1988. Hubble provided this sharpest ever view of the Red Planet (above) by an Earth-based telescope, showing frosty white water ice clouds and swirling orange seasonal dust storms, with a resolution good enough to see details as small as 10 miles (16 km) on the surface. A large dust storm is seen near the north polar cap while another is spilling out of the Hellas impact basin in the southern hemisphere. The distance between Earth and Mars varies considerably, and as a result of the Red Planet's particularly elliptical orbit, at times it can come even closer to the Earth— just 35 million miles (56 million km).

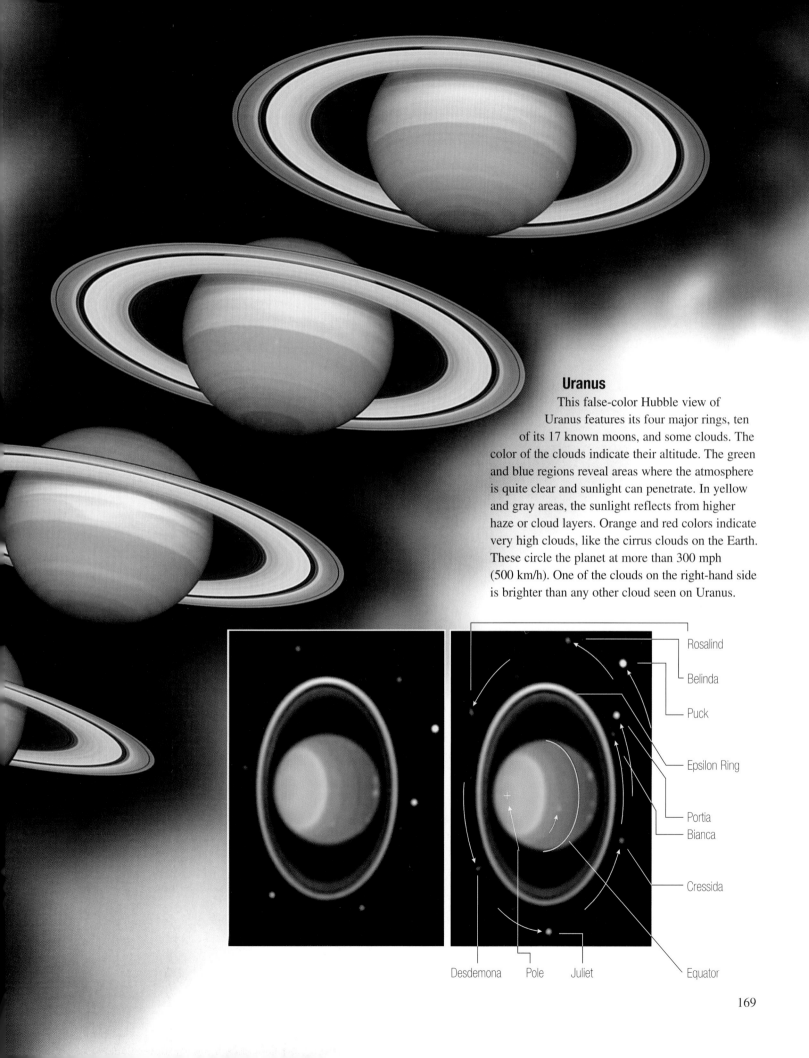

Uranus

This false-color Hubble view of
Uranus features its four major rings, ten
of its 17 known moons, and some clouds. The
color of the clouds indicate their altitude. The green
and blue regions reveal areas where the atmosphere
is quite clear and sunlight can penetrate. In yellow
and gray areas, the sunlight reflects from higher
haze or cloud layers. Orange and red colors indicate
very high clouds, like the cirrus clouds on the Earth.
These circle the planet at more than 300 mph
(500 km/h). One of the clouds on the right-hand side
is brighter than any other cloud seen on Uranus.

Rosalind

Belinda

Puck

Epsilon Ring

Portia

Bianca

Cressida

Desdemona   Pole   Juliet   Equator

Close-ups of the Galaxies

Whirlpool galaxy

What does the Milky Way galaxy look like from deep space? We can get a good idea from this extremely detailed image. Just under the last star on the "bent handle" of the "saucepan" that is the Big Dipper or the constellation of Ursa Major, the Great Bear, is one of the most famous galaxies—M51, the Whirlpool. It is a classic example of a spiral galaxy and is how our Milky Way galaxy might look.

This classic image—also enhanced with

ground-based telescope data—shows the detail of the spiral arms and dust clouds, which are the birth sites of massive and luminous stars. It is a composite image which details visible starlight as well as light from the emission of glowing hydrogen, which is associated with most luminous young stars. Intricate structure is seen for the first time in the dust clouds and along the spiral arms. Dust spurs are seen branching out almost perpendicular to the main spiral arms.

Circinus galaxy

Like a swirling witch's cauldron, this is the galaxy in the constellation of Circinus, 13 million light years away, seen in the southern hemisphere. It is a Seyfert-type galaxy, a class of mostly spirals that have compact centers, and are believed to contain massive black holes. Much of the gas in the disc of the Circinus spiral is concentrated in two specific rings, a large one 1,300 lights years in diameter, and a previously unseen ring, 260 light years in diameter. The rings comprise gas and dust, in which new stars are rapidly forming, apparently in timescales of 40 to 150 million years. At the center of the galaxy is the Seyfert nucleus, the believed signature of a supermassive black hole that is accreting surrounding gas and dust.

Abell galaxy cluster

This giant, cosmic magnifying glass is a massive cluster of galaxies called Abell 2218, in the constellation of Draco, two billion light years away. The cluster is so massive that its enormous gravitational field deflects light rays passing through it to form an image. This phenomenon, called gravitational lensing, magnifies, brightens, and distorts images from faraway objects. The cluster's magnifying powers provide a natural zoom lens, allowing us to view galaxies that we would not normally be able to see, even using the most powerful telescopes.

This phenomenon has produced arc-shaped patterns throughout the Hubble image. These arcs are distorted images of very distant galaxies, which lie 5–10 times further away than the lensing cluster. This distant population existed when the universe was just a quarter of its present age.

The image is dominated by spiral and elliptical galaxies, resembling a string of Christmas tree lights. The biggest and brightest galaxies are members of the foreground cluster.

The Birth and Death of Stars

Thackery

In 1950, astronomer A.D. Thackery peered into a nebula, a star-birth area known as IC2944, and spotted what looked like globules. The area—now known as Thackery's Globules—is filled with gas and dust that is illuminated and heated by a loose cluster of massive stars much larger and hotter than the sun. Hubble provided astronomers with a look at the detail of these globules for the first time, who found that they are in a constant churning motion, moving around each other at high speed. This is the result of powerful ultraviolet radiation from the luminous massive stars, which also heats up the gas in a region of glowing hydrogen, causing it to expand against the globules, which leads to their destruction. They look serene, but according to NASA the globules are like "clumps of butter in a red-hot frying pan."

The Hubble classic

This image of the M16 Eagle nebula, 6,500 light years away in the constellation of Serpens, is considered to be a classic of the Space Age. Finger-like, monstrous columns of cold dust and gas over one light year long protrude from a wall of a vast cloud of molecular hydrogen. The interstellar gas is dense enough to collapse under its own weight, forming young stars. At the tips of the columns are newborn stars emerging from "eggs," or globules, of compact interstellar gas called evaporating gaseous globules. The whole of our solar system would be swallowed up inside the tip of one of the "eggs" at the top of the longest column on the left.

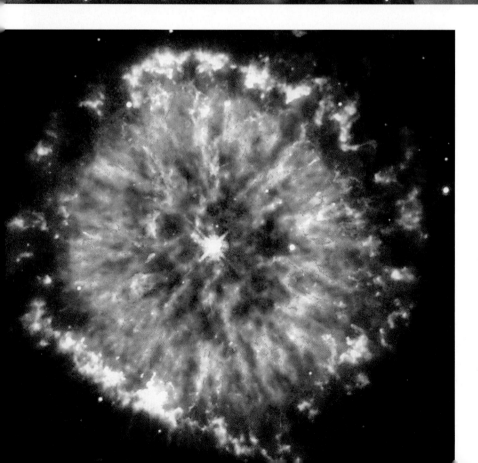

The death of a star

Planetary nebulas are the funeral parlor of the universe. Here is the NGC 6751 planetary nebula in the constellation of Aquila. The nebula is a cloud of gas ejected thousands of years ago from a hot, dying star, still visible in its center. The shell of gas thrown off by the star is its outer gaseous layer, exposing the hot stellar core. The strong ultraviolet radiation then causes the ejected gas to fluoresce as a planetary nebula, a roughly circular ring of radiated gas around the stellar remnant. The vivid colors result from the use of filters used by Hubble's camera. Orange and red show the cooler gas radiating out as streamers, under the influence of radiation and stellar winds from the still hot central dying star, at 140,000° C (252,000°F).

Stellar Varieties

The Pistol Star

Hubble has found one of the most massive stars known, emitting as much as ten million times the power of the sun, and with a radius larger than the distance between the Earth and the sun. It is expelling its outer layers in violent eruptions, producing a nebula four-light years across, surrounding it at a rate of ten masses of the sun in an estimated three million years. It is also one of the most luminous known in our galaxy. The star is destined for a short life and an abrupt end—a spectacular supernova. This near-infrared image shows the erupting star and the gas it expelled showing up prominently as ionized hydrogen gas.

Double cluster

Two dazzling clusters of stars called NGC 1850 are pictured in the Large Magellanic Cloud, one of our neighboring galaxies. The centerpiece is a young, globular-like star cluster of a type that is rare in our Milky Way galaxy. A smaller cluster is below and to the right of the main cluster. The stars are surrounded by a filigree pattern of diffuse gas, seen on the left, which scientists believe was created by an explosion of massive stars.

The Red Giant

This is the first direct image of a star other than the sun. Hubble viewed the famous star, Betelgeuse, in the prominent constellation of Orion. The red supergiant marks a shoulder of The Hunter. The ultraviolet image reveals a huge atmosphere with a hot spot on the surface more than ten times the size of the Earth, suggesting a totally new physical phenomenon that may affect the atmospheres of some stars. Betelgeuse is so huge that if it replaced our sun, its outer atmosphere would extend past the orbit of Jupiter. Hubble can resolve the star even though the apparent size is 20,000 times smaller than the width of the full moon; the equivalent of being able to resolve a car's headlights at a distance of 6,000 miles (10,000 km).

Eyes on the Universe

The Hubble Space Telescope is one of the four NASA spacecraft in the Great Observatory series, which together comprise the biggest single astronomical space program in history. Meanwhile, the European Space Agency has launched the Newton X-Ray telescope as part of the Cornerstone Horizon 2000 science missions, also of great scientific value. NASA's Next Generation Space Telescope, Hubble's replacement, will be launched in 2010. Together, these spacecraft will dramatically increase our knowledge of the cosmos.

Below: The European Space Agency's Newton X-Ray Telescope was launched in 1999 aboard an Ariane 5 booster.

The other telescopes in NASA's Great Observatory series are the Compton Gamma Ray Observatory, which was launched on the Space Shuttle *Atlantis* in 1991, and which is no longer operating, the Chandra X-Ray Observatory, launched in 1999, and the Space Infrared Telescope Facility, which is due to be launched in 2003.

Astronomers observe the universe by studying the electromagnetic spectrum, which includes gamma, X-Ray, ultraviolet, visible light, infrared, radio, and microwave radiation. Visible light, short wavelength radio waves, and some infrared light is able to penetrate the atmosphere and can be observed on the ground, but spacecraft are needed to monitor other wavelengths from outside the atmosphere. Microwave and radio astronomy can be used to examine objects much further away, and radio waves can travel through obstacles such as dust clouds without being distorted.

Ultraviolet observations are useful for looking at material surrounding objects rather than the objects themselves, such as interstellar media. X-Ray astronomy looks at very energetic bodies that emit huge quantities of X-Rays, such as black holes and neutron stars. Gamma rays can provide information on the properties of high energy bodies such as supernovas, pulsars, and quasars. Cosmic rays are the most powerful form of radiation, and can be studied to monitor supernova explosions or nearby solar events.

Visible light images give the world a wonderful view of the universe, as it would be seen by the human eye. Infrared, meanwhile, is essentially heat, and is especially important for looking for cool objects such as dust clouds and dead stars. The far

visible and near infrared region is the one in which astronomers hope to see the components of the universe when they were under development.

Viewing pulsars and black holes

The third Great Observatory series spacecraft was formerly known as the X-Ray Astrophysics Facililty. Renamed the Chandra X-Ray Telescope, it was deployed into orbit by the Space Shuttle *Columbia* in July 1999. The 45 ft (14 m) long spacecraft is equipped with four instruments; an imaging spectrometer, high resolution camera, and high and low energy spectrometers.

Operating in tandem with Chandra is the European Space Agency's (ESA) Newton X-Ray Telescope, which was also launched in 1999. This is providing astronomers with a wealth of data, including images showing a pulsar inside a planetary nebula, and material which seems to be disappearing down a black hole. Newton was formerly called the X-Ray Multi-Mirror Mission (XMM) telescope, and was launched by Ariane 5 in December 1999 as the second of ESA's Cornerstone Horizon 2000 science missions. The craft carries high throughput X-Ray telescopes with an unprecedented effective area, and an optical monitor, the first flown on an X-Ray observatory. The large collecting area and ability to make long uninterrupted exposures provide highly sensitive observations.

NASA has recently launched several other astronomical spacecraft, including the Far Ultraviolet Spectroscopic Explorer (FUSE), which was launched in 1999 to extend the observations of the universe made by previous ultraviolet explorers, with greater sensitivity and resolving power. FUSE will help to answer some fundamental questions about the first few minutes of the Big Bang, by studying the cosmic abundance of deuterium, the heavy hydrogen.

An ESA spacecraft, Infrared Space Observatory (ISO), launched in November 1995, observed the sky with enhanced sensitivity and resolution using a 20 in (51 cm) diameter primary mirror and four science instruments, an infrared camera, photo polarimeter, and two spectrometers. The craft was able to operate for eight months beyond its predicted life, until May 1998, when the coolant was depleted. A short wavelength spectrometer, however, was used until 2001. Another ESA Horizon 2000 mission, the International Gamma Ray Astrophysics Laboratory, Integral, was launched on a Russian Proton booster on October 17, 2002, into a high 24,800 miles (40,000 km) Earth orbit.

NASA's Next Generation Space Telescope, renamed the James Webb Space Telescope in honor of the man who led NASA during the early days of the Space Age, will be launched in 2010. It will take a further three months to reach a point 940,000 miles (1.5 million km) from Earth, where the spacecraft will be equally balanced between the gravity of the Earth and the sun. This telescope will peer even further into the universe than Hubble could.

Exploring the Nearest Star

Despite the contributions made by Earth-based telescopes, we have learned much more about the sun's activity using spacecraft. The ESA-NASA Solar and Heliospheric Observatory (SOHO), and NASA's Transition Region and Coronal Explorer (TRACE), are returning spectacular images of our highly active and turbulent star. These and other craft provide daily "weather" reports of the sun and its effect on the Earth's spatial environment.

Below: The European Space Agency's Solar and Heliospheric Observatory (SOHO) has conducted the most detailed continuous observation of the sun in history.

Our sun is small compared to many other stars, and is categorized as a yellow dwarf, but is still 109 times bigger than the Earth. The sun is a turbulent sphere of very hot gases produced by continuous nuclear reaction or fusion. Every second, 700 million tons of hydrogen fuse together to form helium, releasing energy in the form of heat and light.

The sun may look tranquil when seen in visible light, but images in other wavelengths taken by SOHO, TRACE, and other spacecraft, reveal bursts of giant, spectacular flames and storms. Sunlight takes eight minutes and 17 seconds to reach the Earth, which is 93 million miles (149 million km) away. The sun has a diameter of 860,000 miles (1.4 million km), and its surface temperature ranges from 4,300° C (7,800°F) to 9,000°C (16,200°F).

Storms on the surface, the photosphere, create sunspots. These violent storms emit solar flares. The upper level of the photosphere is called the chromosphere, which has very hot gases with temperatures up to one million°C (1.8 million°F). Above the chromosphere, the sun has a corona, a halo of even hotter gases as high as 4 million°C (7.2 million°F). The sun emits radiation called the solar wind, and its charged particles become trapped in the Earth's magnetic field and interact with gases in the upper atmosphere. This creates the aurora borealis in the north, and the aurora australis in the south.

Continuous monitoring

All this activity and its effect on the Earth is being closely watched by satellites, and by spacecraft in interplanetary space, such as SOHO and the four ESA Cluster satellites, working in concert as part of the Solar Terrestrial Science program.

SOHO was launched in December 1995. Its mission is to investigate the processes that form and heat the corona, maintain it, and give rise to the expanding solar wind. Also investigating the internal structure of the sun, SOHO carries 11 instruments including a range of spectrometers and particle analyzers. The craft is situated in a special orbit between the sun and the Earth, 940,000 miles (1.5 million km) from the Earth, where it points to the sun continuously, with the Earth behind the spacecraft.

The four Clusters were orbited in 2000 to study the three-dimensional extent and dynamic behavior of Earth's plasma environment, observing how solar particles interact with the Earth's

magnetic field. Passing in and out of the Earth's magnetic field, the satellites observe such phenomena as the magnetopause, bow shock, and magnetotail.

Meanwhile, the U.S.'s TRACE was launched in April 1998 to perform the first American mission dedicated to solar science since the Solar Maximum Mission satellite was launched in 1980. TRACE takes high-resolution images of the transition region between the sun's photosphere and the corona, to obtain measurements of the temperature regimes and to complement data being collected by other spacecraft including SOHO.

Several solar orbiting spacecraft, including those launched very early in the Space Age, sent back data about the sun. More recently, the first spacecraft to orbit the sun around its poles, out of the ecliptic plane in which the planets circulate, was *Ulysses*. It was launched in October 1990, and passed over the south solar pole in September 1994. Equipped with a suite of esoteric solar study experiments, *Ulysses* flew over the north solar pole in July 1995, and repeated a pass over the south pole in January 2001.

Above: NASA's Transition Region and Coronal Explorer took this image of the sun's photosphere, with the Earth placed alongside for scale.

Left: ESA's *Ulysses* was the first spacecraft to fly over the poles of the sun, in 1994–95.

Searching for the Beginning

In recent years, many fundamental questions about the Universe have been answered by large telescopes, especially Hubble. As is often the case, however, some of the findings have thrown up more questions than they have answered. Scientists are tantalizingly close to being able to see the processes which went on in the very early Universe. It is hoped that even more powerful telescopes will soon answer the biggest question of all—the age and origin of the Universe.

NASA's Space Infrared Telescope Facility (SIRTF) is the last of the Great Observatory series spacecraft, after Hubble, Compton, and Chandra. It will be launched in January 2003, and is expected to be named after a well-known infrared astronomer. SIRTF is much smaller than originally planned, and will be placed into an Earth-trailing solar orbit, which will provide excellent visibility and a benign thermal environment. Its liquid helium-cooled telescope is expected to operate for about three years before the cryogenic coolant is depleted.

The next big space telescope project will be the $2 billion Next Generation Space Telescope (NGST), a 24 ft (8 m) instrument combining visible and infrared astronomy. It will be placed into an orbit between the Earth and the sun in 2010,

940,000 miles (1.5 million km) from Earth. The telescope will be a key component of NASA's new Origins program, helping to answer key questions about the nature of the Universe, including its creation, physics, chemistry, and potential biology, and to find out whether there are Earth-like planets in existence.

NGST will be able to observe first generation stars and galaxies, individual stars in nearby galaxies, penetrate dust clouds, and discover thousands of objects in the Kuiper Belt. The new telescope will be able to see objects 400 times fainter than those currently studied by the largest ground-based telescopes or by space-based infrared satellites, with a sharpness surpassing that of the Hubble Space Telescope. The NGST will be optimized for infrared because the expansion of the universe is causing the light of the most distant objects to be shifted into the infrared spectrum.

Cosmic mystery

Astronomers are still struggling to work out the age of the universe, with estimates of 15 billion years being the most popular. Current theories about the universe's age are under scrutiny, however, since images of very distant objects—the early cosmos—

Right: The final NASA Great Observatory, the Space Infrared Telescope Facility (SIRTF) will be launched in 2003.

Opposite: The successor to the Hubble Space Telescope will be the Next Generation Space Telescope, which will have to make a quantum leap in technology to improve on Hubble 's resolution.

taken by Hubble and
other telescopes, do not
show the early stages of
galaxy formation that scientists
expected. It is hoped that the NGST
will, quite literally, shed light on this
discrepancy. One spectacular image taken
by Hubble in 1996 was called the "Beginning
of Time." It shows perfectly formed galaxies
12 billion years old, puzzling scientists, who had
expected to see galaxy development at work.

It is now thought that the earliest galaxies may
have formed within only about a billion years of
the Big Bang. Recent HST images, for example,
show what appear to be galactic building blocks
within a few hundred million years of the Big
Bang. Will we ever see images of galaxy
formation? The NGST should answer that question.

NASA also wants to help astronomers
understand the shape of the universe, the interaction
of stars and planetary systems, the life cycles of
matter in the universe,
and to find out about the dark
matter in the cosmos.
A final objective of the Origins program
is to answer "are we alone," and the
NGST is expected to find planetary systems
around stars, with the possibility that an
Earth-like planet could be discovered.

CHAPTER THIRTEEN

INTO THE FUTURE

The Space Traffic-jam

One of the first man-made objects to orbit the Earth was a piece of space debris. Designated 1957 alpha, it was the conical cover that protected the satellite, Sputnik 1, which was designated 1957 beta. The spent rocket stage was designated 1957 gamma. The designation method has changed since then, but the problem of space debris has not.

O rbital space debris is a big problem, and it will not go away. Today, almost 9,000 man-made objects larger than a baseball—of which only about 700 are operational satellites—can be tracked in Earth orbit.

Of these, 41% are large fragments left over from explosions, usually caused by unspent propellant on rocket stages. Discarded upper stages represent 17% of debris; defunct satellites, 22%;

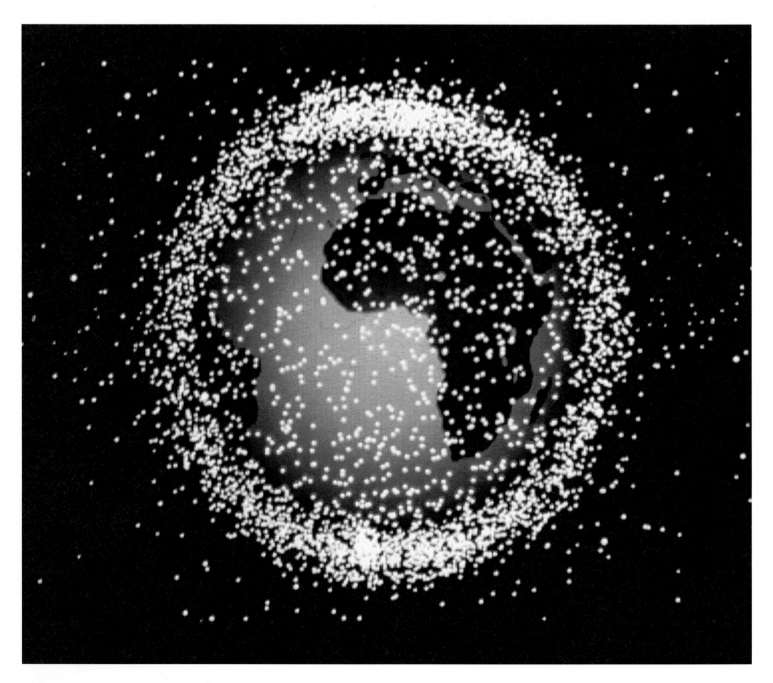

impacts were caused by tiny aluminum oxide particles from solid rocket motors. A tiny fragment could easily puncture an EVA spacesuit. A window of the Space Shuttle *Challenger* was chipped by a 0.3 mm fleck of paint which impacted at four miles per second.

New space debris problems are now appearing—and the perennial ones are repeating themselves. A Russian electronic intelligence satellite, *Cosmos 2347*, which had reached the end of its operational life, was deliberately self-destructed in orbit, creating over 130 pieces of more debris in a 145 mile (230 km) by 255 mile (410 km) orbit. The self-destruct policy is now a rare one and is a throwback to the security-conscious times of the Cold War.

The Space Shuttle has made several evasive maneuvers during missions, to keep a wide berth of even the smaller class of debris, which is being carefully tracked by Norad in Colorado. Debris is tracked worldwide in numerous ways, from radar to high-resolution optical telescope cameras. This system is far from perfect, however. Recent radar measurements of debris show the ratio of cataloged to uncataloged objects larger than 4 inches (10 cm) as about 1:4.

Optical measurements taken in circular geostationary orbit, 22,000 miles (36,000 km) above the equator (the operational position of hundreds of communications satellites), show a large number of previously undetected objects of 8 inches (20 cm) to 40 inches (100 cm) size.

Not all debris stays in space, however. Parts of upper stages that were previously believed to have burned up completely during re-entry have been found intact on the ground, adding new concern to another problem—the potential for space debris to strike people and property on Earth.

Many measures have been suggested by various international bodies, but until a major accident occurs, it is unlikely that anything will be done about space debris.

and ejected payload shrouds and covers, tools left over by spacewalking astronauts, and other items, 13%. To make matters even worse, it is estimated that over 150,000 pieces of debris smaller than the size of a tennis ball and down to half an inch (1 cm) diameter are in orbit.

Their number is increasing, despite a relative reduction in the annual launch rate. The potential for disaster is obvious. A half-inch-size fragment orbiting at speeds of over 17,340 mph (28,000 km/h), or 5 miles per second (8 km/s), could shatter a $100 million satellite—or indeed the Space Shuttle or an International Space Station module.

Crowded sky

France's satellite *Cerise*'s long antenna was severed by a fragment from an Ariane 4 third stage—at a relative speed of 12 miles per second (20 km/s)—in the first known collision of space objects, in 1996. Spacecraft are also being peppered in orbit by even smaller pieces of debris every day.

After 68 months in space, the NASA *Long Duration Exposure Facility* was recovered and inspected. It had been hit by 34,000 micron-size particles (a thousandth of a millimeter), the largest of which made a 5.25 mm diameter crater. Some

Facing page: The European Space Agency's depiction of the population of space debris in low Earth orbit.

Left: Inspection of NASA's Long Duration Exposure Facility, after years in orbit, gave an indication of the damage that debris causes to spacecraft.

Above: Spacewalking astronauts are most in danger from debris. A tiny particle impacting a suit at thousands of miles an hour could be fatal.

Onward and Outward

Interstellar travel is a reality—but only just. Four NASA spacecraft, Pioneers 10 and 11 and Voyagers 1 and 2, have left the solar system and are heading for the stars. Pioneer 10, launched in March 1972, was for 30 years the most remote man-made object, and is now about 7.5 billion miles (12.07 million km) away. It has since, however, been overtaken by Voyager 1, which by March 2002 was 7.7 billion miles (12.4 billion km) distant.

Pioneer 10, which was the first spacecraft to explore Jupiter, made valuable scientific investigations in the outer regions of our solar system until the end of its science mission on March 31, 1997. However, its weak signal continues to be tracked by NASA's Deep Space Network antennas, as part of an advanced study of communications technology in support of a possible NASA interstellar probe mission.

Pioneer 10 will continue into interstellar space at a speed of 7 miles per second (12 km/s), heading in the general direction of the red star Aldebaran, which forms the eye of the constellation of Taurus, the Bull. Aldebaran is about 68 light years away, and it will take Pioneer 10 over two million years to reach it.

The Pioneer 11 mission ended officially on September 30, 1995, when the last transmission from the spacecraft was received. It was launched to Jupiter and Saturn in April 1973. The spacecraft is heading toward the constellation of Aquila, the Eagle, and will pass near one of the stars in the constellation in about four million years.

Below: "We are here" says a message on a plaque on Pioneer 10, showing the solar system and flight path of the craft, together with two humans.

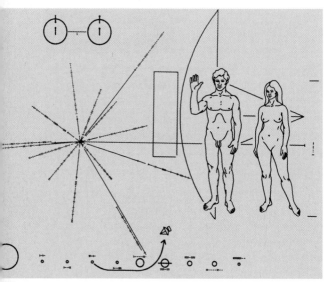

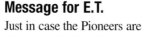

Message for E.T.

Just in case the Pioneers are discovered by intelligent beings far out in space, they each carry a plaque. The 5.9 inch (15 cm) by 9 in (23 cm) gold-plated aluminum plaque is etched with a diagram showing male and female human beings, with their heights relative to the size of the spacecraft. The male is raising his hand in greeting. The path of the spacecraft across the solar system is shown in relation to the position of 14 pulsars.

Rechristened the Voyager Interstellar Mission (VIM) by NASA in 1989, after Voyager 2's Neptune encounter, Voyagers 1 and 2 continue operations, taking measurements of the interplanetary magnetic field, plasma, and charged particle environment. The spacecraft also search for the heliopause, the place at the edge of the solar system where the solar wind becomes subsumed by the more general interstellar wind.

By the end of the Neptune phase of the Voyager project, a total of $875 million had been expended for the construction, launch, and operations of both Voyager spacecraft. An additional $30 million was allocated for the first two years of VIM. Voyager 1 is now the most distant man-made object in space, 7.7 billion miles (12.4 billion km) from the Earth.

It is heading to within three light-years of a dwarf star in the constellation of Camelopardalis, the Giraffe, in 400,000 years time. Voyager 1 was launched in September 1977 and explored Jupiter and Saturn.

Voyager 2, launched in August 1977 (before Voyager 1), also explored Jupiter and Saturn, as well as Uranus and Neptune, and is now 6.3 billion miles (10.1 billion km) from the Earth. It is heading for a 0.8 light-year distance of Sirius—the brightest star in our skies—in about 358,000 years time.

Each Voyager carries a sample of Earth life and culture in case it is intercepted. A 11.8 inch (30 cm) diameter gold-plated copper phonograph disc (together with a needle and playing instructions!) bears sounds from Earth, including thunder, bird songs, whales communicating, volcanoes, and even human laughter. The disc also contains 90 minutes of music and 115 analog pictures, and greetings in 60 languages.

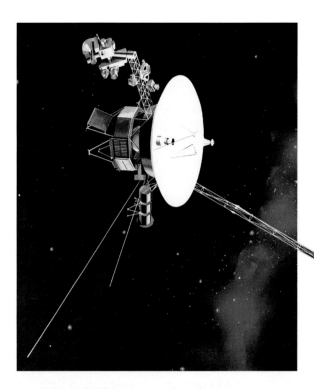

Voyagers 1 and 2 (spacecraft, **left**) and Pioneers 10 and 11 (spacecraft, **below**) are heading into deep space. Voyager 1 will be the first to pass near a star, Sirius, 358,000 years from now.

Down to Earth

We are currently less than half a century into the Space Age. If we compare the Space Age to the development of aviation, we would find that we have only reached the equivalent of the 1940s. The rocket engine of today is basically the same as it was in 1957. Who knows what leaps and bounds will be made in space technology during the next half century?

Two perfectly rational predictions at the time of Apollo were that humans would land on Mars in 1985, and that the President of the United States would travel to a space station in 2000 in a space taxi.

If the U.S. had allocated 20 times the amount it spent on Apollo, it may well have reached Mars by now. As events turned out, it eventually spent five times Apollo's money on the Space Shuttle. If a space station and the Space Shuttle had been developed as an integral part of a stepping stone to a Mars landing, then we may have been closer to achieving that aim.

Space travel is currently extremely expensive, and the first task in the 21st century should be to make it much cheaper. Access to space will have to be as routine as air travel—and as safe—if great advances are going to be made. However, further progress in space travel is likely to be slow, judging by NASA's admission that the Space Shuttle could still be flying in 2020—or later. Cost, routine access, and safety will be dominant factors in the future of human space travel, and quantum leaps in technology will have to be made before we move into a new space era.

Above: A 1980s depiction of a new generation of space station, serving as a staging post for flights to the moon and planets.

Right: In the late 1950s and early 1960s, an abundance of concept art was published to show us how exciting life on the moon would be in only a few years. Alas, it has not come to pass—we are left with yesterday's dreams. Perhaps this *is* how a moon base could look *if* we decide to return to our nearest neighbor.

Attempting to predict the future of space exploitation and exploration is fraught with difficulty, partly due to the fact that it is difficult to distinguish media hype and publicity from true progress. NASA and other space organizations often hype-up potential missions in an effort to inspire the public and obtain finances, but then the whole scheme is buried. This is particularly the case with future launcher projects, such as the fully reusable Space Shuttle successor vehicle, the legendary VentureStar, which was actually no more than a project on paper, and never survived further than the drawing board. A Mars sample return mission is also unlikely before 2015, and in fact terrestrial priorities may postpone such a mission even longer.

Courage and a will

However, it is not all doom and gloom. New space technologies, such as the development of artificial intelligence for use on spacecraft, and new ways to propel craft through space much faster, are being developed. It is indeed likely that space applications, such as communications, navigation, and the development of new propulsion systems, will advance rapidly.

As for manned flights to other planets, these are still the realm of science fiction, and are unlikely before 2020 at the very earliest, unless that unique combination of factors that gave birth to Apollo—technological development, political will, and competition—occur again.

Human space travel in the future will depend a great deal on safety. However, it must be accepted that accidents will happen. After *Columbia*, the loss of yet another Space Shuttle, or any manned space vehicle, however tragic an event that would be, must be accepted as probable. Just as Yuri Gagarin and John Glenn risked their lives in the early days of the Space Age, so the astronauts and cosmonauts of the future will need to accept risk as an inevitable precedent to progress.

Below: The ultimate goal—the manned exploration of Mars and its moons. The first simple, human expedition to Mars is unlikely until 2020.

"In the beginning, God created the heavens and the Earth..."

Extract from *Genesis*, read aboard Apollo 8 by crew members Anders, Lovell, and Borman, December 1968.

Arthur C. Clarke has said that the future of space travel will be motivated by applications. Manned Mars landings and travel to the stars are not unrealistic dreams in the long-term, he said, but the immediate future of space technology will be dictated by more immediate practical uses.

Clarke came to the fore in 1945, when he predicted that TV and communications satellites would soon be placed in geostationary orbit. In 1951, he also predicted that satellites would be used to "send mail"—another astonishingly accurate statement considering the recent development of e-mail.

Asked recently why he thought a mission to Mars had not occurred, Clarke said it was due to "Vietnam, Watergate, and the ending of the Cold War. The space program was politically motivated." The political raison d'être has indeed gone. *2001: A Space Odyssey* has remained a dream largely due to economics and politics, rather than technology. Clarke added his opinion that even if we go back to the moon and land on Mars in 2020 or 2030, it will be technology-led, and therefore a slower program than the politically-motivated Apollo rush to the moon.

Getting Neil Armstrong onto the Sea of Tranquillity within eight years was an extraordinary achievement, and an illustration of what man can do if he wants to—money permitting. Today, it is a case of "when can we do it?" rather than "it must be done." The future of space travel will depend largely on the development of new propulsion and life-support systems. However, Arthur C. Clarke maintains that, with the necessary will, it would be technologically feasible to "go to the moon for the same as it costs to fly around the world."

Carl Sagan wrote that "in the long term, every civilization must be spacefaring, otherwise it will die." Joking, Sagan once famously said that the dinosaurs became extinct because they did not have a space program! (The extinction of the dinosaurs is thought to have been caused by the impact of a huge comet or asteroid with the Earth).

On the same theme, Clarke has said that "the danger of an asteroid, or cometary impact, is the biggest single reason now for developing space technology, so we can destroy or divert an approaching object."

"Things have gone beyond anything I ever imagined," he said. Once he didn't believe he would live to see men go to the moon, but he never believed that "we would go there and abandon it after five years."

Right: Earthrise, as seen by Apollo 11 in lunar orbit in July 1969.

Our place in the universe

What if we are alone? "Then we are heirs to the cosmos, and its guardian," said Clarke. One of the main goals of space exploration today is to prove that we are not alone or just an accident in the Universe. Mankind has a deep need to find out the answer to this question.

Albert Einstein warned that we need godly humility, something that has been all but lost in the scientific world today. When he wrote those words, more people believed in the Biblical creation than in evolution; now it seems to be the other way around.

The irony of space exploration is that, in going out into space, to the moon and the planets, with dreams of exploring the stars—going "where no man has gone before"—we have come back with an image far more beautiful than a picture of a dark moon crater or a Martian rock. We have seen the planet Earth from space.

The photograph of the Earth, a fragile planet with an extraordinary range of life, in the vastness and blackness of space, is one we are now all familiar with. It reminds us that the best place in the Universe that we currently know of is right here.

Tim Furniss

Milestones in Space Exploration

March 16, 1926
World's first liquid propellant rocket launched by Robert Goddard in USA

November 25, 1933
Russian GIRD 10 liquid rocket launched

October 3, 1942
German A-4 rocket (V-2 precursor) launched to 50 miles altitude

April 15, 1946
First U.S.-assembled V-2 launched from White Sands

October 24, 1946
U.S. V-2 flies to 165 miles altitude

February 24, 1949
U.S. Bumper WAC V-2-Corporal flies to 242 miles

June 14, 1949
Monkey Albert 2 flew to 83 miles by V-2

May 24, 1954
U.S. Viking rocket reaches 158 miles altitude after launch from White Sands

August 27, 1957
Soviet Union R-7 ICBM flies 5,000 miles in distance

October 4, 1957
R-7 derivative launches first satellite, Sputnik 1

November 3, 1957
Soviet Sputnik 2 carries the dog Laika, who was not recovered from orbit

January 31, 1958
Explorer 1, America's first satellite launched

May 28, 1959
U.S. monkeys Able and Baker recovered after 300 miles altitude flight

September 13, 1959
Soviet Luna 2 becomes first spacecraft to hit the moon

October 6, 1959
Soviet Luna 3 sends back images of 70% of the moon's far side

April 1, 1960
Tiros 1, the first weather satellite, launched by USA

August 11, 1960
U.S. Discoverer 13 capsule is first object to be recovered from orbit

April 12, 1961
First man in space is Yuri Gagarin, launched by USSR aboard Vostok 1

May 5, 1961
America's first astronaut, Alan Shepard, is launched on sub-orbital Mercury flight

February 20, 1962
John Glenn becomes first American in orbit aboard *Friendship 7*

July 10, 1962
Telstar 1, the first TV communications satellite, is launched by U.S.

December 14, 1962
U.S. Mariner 2 makes first exploration of Venus

June 16, 1963
Soviet Valentina Tereshkova becomes first woman in space, aboard Vostok 6

July 31, 1964
U.S. Ranger 7 impacts moon after taking first close-up images

October 12, 1964
Soviet Voskhod 1 carries 3 cosmonauts

March 18, 1965
Alexei Leonov becomes first person to walk in space, from Voskhod 2

June 3, 1965
Edward White becomes first U.S. spacewalker from Gemini 4

July 15, 1965
U.S. Mariner 4 explores Mars for first time during fly-by

December 16, 1965
First rendezvous in space performed by Gemini 6 astronauts

February 3, 1966
Soviet Luna 9 makes "soft-landing" on moon

March 16, 1966
First docking in space made by U.S. Gemini 8 crew

March 31, 1966
Soviet Luna 10 is first lunar orbiter

January 27, 1967
Apollo 1 fire kills three astronauts during ground test

April 24, 1967
Soviet Vladimir Komarov is first space casualty in crash of Soyuz 1

December 21, 1968
U.S. Apollo 8 carries first humans to orbit the moon ten times

July 20, 1969
U.S. Apollo 11 is first manned landing on moon. First walk by Neil Armstrong

December 15, 1970
Soviet Venera 7, first landing on Venus

April 19, 1971
Soviets launch first Salyut space station,

June 29, 1971
Soyuz 11 crew fly for 23 days but killed when craft depressurizes on return

November 14, 1971
U.S. Mariner 9 is first Mars orbiter

December 7, 1972
U.S. Apollo 17 is launched on final moon landing mission

May 14, 1973
U.S. Skylab 1 space station launched. Skylab 3 crew flew 84-day mission

December 5, 1973
U.S. Pioneer 10 makes first exploration of Jupiter

March 29, 1974
U.S. Mariner 10 makes first exploration of Mercury

June 8, 1975
Launch of Soviet Venera 9, the first Venus orbiter

July 20, 1976
U.S. Viking 1 makes first landing on Mars

June 15, 1979
Soviet Soyuz 29 crew launched on 139-day mission on Salyut 6

September 1, 1979
U.S. Pioneer 11 makes first exploration of Saturn

April 12, 1981
The first U.S. Space Shuttle launch, of orbiter *Columbia*, the first reusable spacecraft

May 13, 1982
Soyuz T-5 crew launched on 211-day mission to Salyut 7

February 3, 1984
Bruce McCandless launched on Space Shuttle mission to fly first Manned Maneuvering Unit (MMU) on EVA

November 8, 1984
Space Shuttle launched on first satellite retrieval and return to Earth mission

January 24, 1986
U.S. Voyager 2 makes first exploration of Uranus

January 28, 1986
Seven-person crew of Shuttle *Challenger* killed in launch explosion

February 20, 1986
Soviet Mir space station launched on 15-year mission with multiple crews

March 14, 1986
UK *Giotto* spacecraft flies through the nucleus of Halley's Comet

February 5, 1987
Yuri Romanenko begins 326-day mission on Mir

December 21, 1987
Vladimir Titov and Musa Manarov begin 365-day Mir mission

August 25, 1989
U.S. Voyager 2 makes first exploration of Neptune

April 24, 1990
Hubble Space Telescope launched

October 29, 1991
U.S. *Galileo* makes first exploration of an asteroid, Gaspra

January 8, 1994
Russian Valeri Poliakov launched on Mir mission lasting a record 437 days

June 27, 1995
Space Shuttle launched with U.S./Russian crew for first joint Mir mission

December 7, 1995
U.S. *Galileo* becomes first Jupiter orbiter

February 17, 1997
Launch of U.S. NEAR craft, the first to orbit an asteroid, Eros, and to land on it

July 4, 1997
U.S. Mars Pathfinder deploys first Mars rover, Sojourner, on surface

October 29, 1998
Pioneer astronaut John Glenn, age 77, flies on Space Shuttle

November 28, 1998
First launch of International Space Station module, the Russian *Zarya*

February 1, 2003
Seven-person crew of Shuttle *Columbia* killed during re-entry